UNIVERSAL PATTERNS

Second Revised Edition

The Golden Relationship:
Art, Math & Nature

book one

Second Revised Edition

UNIVERSAL
PATTERNS

by
Rochelle Newman & Martha Boles
(This is a commutative operation)

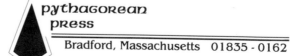
pythagorean
press
Bradford, Massachusetts 01835 - 0162

Second Revised Edition
Copyright © 1992, Pythagorean Press
First Revised Edition, 1990
Second Edition, 1987
First Edition, 1983

All new hand illustrations by Susan Libby, Newburyport, Massachusetts.

All photographs, unless otherwise indicated, are the work of Richard Newman.

ISBN 0-9614504-4-4

This book was illustrated and designed on MacIntosh computers and proofed on a Laserwriter IINT printer. It was printed on recycled paper.

Library of Congress Catalog Card Number: 85-60105

Cover photo: Richard Newman
Summer Scene

About the authors:
Rochelle Newman is an artist and Professor of Art at Northern Essex Community College, Haverhill, Massachusetts.
Martha Boles is Professor of Mathematics at Bradford College, Bradford, Massachusetts.

This text is one aspect of an ongoing collaboration dedicated to the concept of interdisciplinary learning.

1% of all sales is donated for the funding of environmental protection projects.

Morris Kline once said to a group of mathematics teachers, "96.6% of the people are different from us. They don't like mathematics." It is to this large segment of the population that we dedicate this book.

Acknowledgements

In this second revision of Book I, we continue to acknowledge the primary forces of passion, patience, perseverance, cooperation, love and commitment to quality. This book is a manifestation of a synergetic relationship between the authors.

We thank *Bradford College* and *Northern Essex Community College* for allowing us to teach a course that bridges disciplines, institutions and diverse philosophies. Teaching is dialogue, and it is through the process of engaging students that we see ideas taken from the abstract and played out in concrete visual form. Students teach us about creativity through their personal responses to the limits we set, thus proving that reason and intuition are not antithetical. Their works give aesthetic visibility to mathematical ideas. We wish to thank all those students who caught our enthusiasm, coupled it with hard work, and created lovely art works, some of which are shared in this book.

In addition we thank:

Donna Fowler, for her design skills in restructuring the format of this book, and for her work in developing many of the computer generated constructions and diagrams;

Richard Newman, for his art and photography;

Susan Libby, for her hand drawn illustrations;

Sylvia Burnside, for all of her drawings and diagrams that have been carried over from previous editions, and for her hard work during the initial stages of this project;

and our assistants:

Linda Maddox, who created so many excellent drawings with only her technical pen and black ink and lots of skill, and

Ann MacLean, for her fine editorial and proofreading skills.

Contents

Constructions

Preface

It is this silent swerving from accuracy by an inch that is the uncanny element in everything. It seems a sort of secret treason in the universe...
Everywhere in things there is this element of the quiet and incalculable.

The real trouble with this world of ours is not that it is an unreasonable world, nor even that it is a reasonable one. The commonest kind of trouble is that it is nearly reasonable, but not quite. It looks just a little more mathematical and regular than it is; its exactitude is obvious, but its inexactitude is hidden; its wildness lies in wait.

Gilbert Chesteron

To search for relationships between things apparently disconnected is the Aristotelian Quest. Our desire is a variation on this quest: to show relationships between things which appear disconnected. The quest is a way to identify wholeness.

This world is of a single piece; yet, we invent nets to trap it for our inspection. Then we mistake our nets for the reality of the piece. In these nets we catch the fishes of the intellect but the sea of wholeness forever eludes our grasp. So, we forget our original intent and then mistake the nets for the sea.

Three of these nets we have named Nature, Mathematics and Art. We conclude they are different because we call them by different names. Thus, they are apt to remain forever separated with nothing bonding them together. It is not the nets that are at fault but rather our misunderstanding of their function as nets. They do catch the fishes but never the sea, and it is the sea that we ultimately desire.

Our aim in this book is to explore some of the connections; to look at Nature and then to see how the disciplines of Art and Mathematics catch the same fishes of this natural world. It is both the holes in the net and the net itself that we would like to examine. Our concern is for relationships, connections, interfaces, overlaps.

One of us comes from the discipline of Mathematics with an interest in Art; the other comes from the arena of Art with a curiosity about Mathematics. Our complementary skills show that both Art and Mathematics

are necessary for a fuller understanding and enjoyment of the world which we call Nature.

Though Nature is both complex and simple simultaneously, we believe she is comprehensible to the human mind and consistent with her laws. She suggests ultimate order, which is perfection, on one hand; then offers us deviation and disorder on the other. She is ever involved in ceaseless activity. She never gives us the mathematician's precision but rather offers us the artist's abundance and variety. She seldom works with a simple one-force system, but, rather, delights in the interplay of several forces. She must be understood by viewing the interaction of systems synergetically; the whole always being greater than the sum of the parts.

crete, isolated, independently existing bits, but must look at these bits in process and in relation to one another. This planet, this *Spaceship Earth,* this ecosphere can sustain, maintain, generate life and adapt to changing conditions as do the individual organisms that are a part of this system. It is our responsibility to understand this dynamism in order to work in consonance with it. Then, perhaps our arrogance shall give way to the wisdom that we must be responsible participants who are just part of the larger processes.

Despite all our attempts at explanations, the essential enigma will remain. For whenever we use one of our nets to catch a part of Nature we necessarily exclude other parts. Therefore, none of our explanations can be total Truth but rather only a partial truth limited to a particular time and place. As Jacob Bronowski said, "No formal system embraces all the questions that can be asked."

Our senses give us access to only a limited range of Nature's reality. Beyond and below the boundaries of our senses are other aspects of the same reality, but they are imperceptible at certain levels. The world of the fly is not the same as that of the elephant. But at the most minute level there is similarity. We are just now beginning to understand that ours is a

> There is individuality; there is community,
> There is microcosm; there is macrocosm,
> There is visible; there is invisible,
> There is randomness; there is organization,
> There is reason; there is intuition,
> There is mind; there is body,
> There is inside; there is outside,
> There is motion; there is rest,
> There is equilibrium; there is fluctuation,
> There is competition; there is cooperation,
> There is chance; there is choice,
> There is uncertainty; there is predictability,
> There is perfection; there is flaw,
> There is process; there is product,
> There is symmetry; there is asymmetry,
> There is theme; there is variation,
> There is diversity; there is unity,
> There is you; there is me.

Nature is all inclusive; however, for purposes of discussion let us call Nature that which is given, not the man-made nor the artificial. Nature is a verb: partaking of, creating, re-creating the universal interconnectedness and interdependence of all phenomena in an ever dynamic cosmic dance. Each individual dancer is part of an environment which, in turn, is part of a larger environment which is part of a planetary system which, in turn, is part of a solar system which itself is part of a galaxy which is part of the Cosmos. Thus, the dancer and the dance are one. We cannot decompose Nature into dis-modular world built out of multiple replication of the structures which we call atoms. These atoms, which are mostly composed of space, contain other marvelous elements called electrons, protons, neutrons, quarks, ever moving as individual members, but organized in their patterns. All atoms have the same basic internal structure. What makes them different from each other is the number of electrons and protons contained within them. There are only a limited number of elements, each with its particular atomic structure, which become the building blocks of our natural world. Atoms join together to form molecules, which, in turn, cluster to form larger units of matter. The variety that Nature displays is generated by permutations of these basic building blocks.

Despite the essential difficulties in understanding the whole of Nature, both Mathematics and Art reflect a human desire to comprehend aspects of the world. Both come after the fact of Nature and each never fully explains her, but in their complementarity they work to give wholeness.

Both Art and Mathematics are *games*, constructs of the mind, set within recognizable limits. And both disciplines look for relationships among the game pieces. Depending upon the cultural milieu and the problem in need of solution, there is

Below:
Richard Newman. Cosmic Game Board. *Photographic Collage.*

an initial premise as to how the game must be played. The game board must be constructed according to the object of the game and the rules and techniques necessary for playing.

To understand relationships is to search for patterns that are identifiable and repeatable. In Mathematics, however, one interest is in relationships without reference to things. There is no concern for the affect that an object might have upon its perceiver, while in Art the interest in the relationship is *because* of the things. The "thingness" of the object is as strong a concern as its connections. It is, in part, the emotional qualities that engage the artist. If it is the aim of Mathematics to seek the simplest, but all-encompassing, explanations of the most complex of facts, is it not then the aim of Art to express the joy in experiencing the multitude of relationships?

Art looks to the specific to speak about the general; Mathematics uses the general to explain the specific. One is idiosyncratic, the other, impersonal. As the scientist D'Arcy Thompson pointed out, mathematically we can define the form of a wave or a heap of sand, but never ask a mathematician to define the form of any particular wave of the sea, nor the actual form of any mountain peak or hill. That is the task of the artist. Not only will he or she give you the particular wave or mountain peak, but it will be a reflection of his or her own perceptions of that wave or peak, unlike any other.

In this world of overspecialization, much of education deals with discrete bits of information rather than large systems. People, therefore, are not trained to find connections. Without connections, value systems are difficult to develop. In the evolution of civilization, Art and Mathematics are disciplines that have been seen as polarities without connection. Yet, in fact, they are the left and right hand of cultural advance: one is the realm of metaphor, the other, the realm of logic. Our humanness depends upon a place for the fusion of fact and fancy, emotion and reason. Their union allows the human spirit freedom.

Below:
Student works. Pen and ink on paper. Chapter 1 Project 2.
Top left:
Linda Lazzaro.
Bottom left:
Francisco Colom.
Right:
Julie Conway.

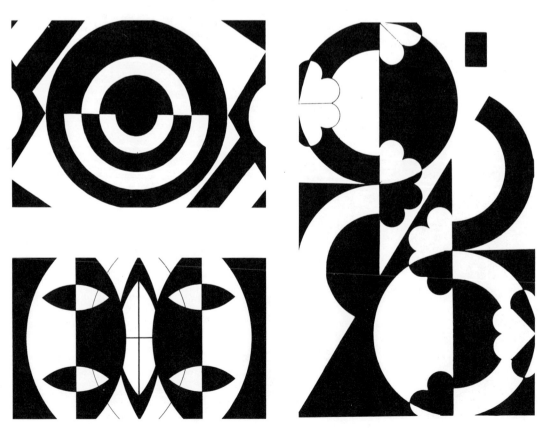

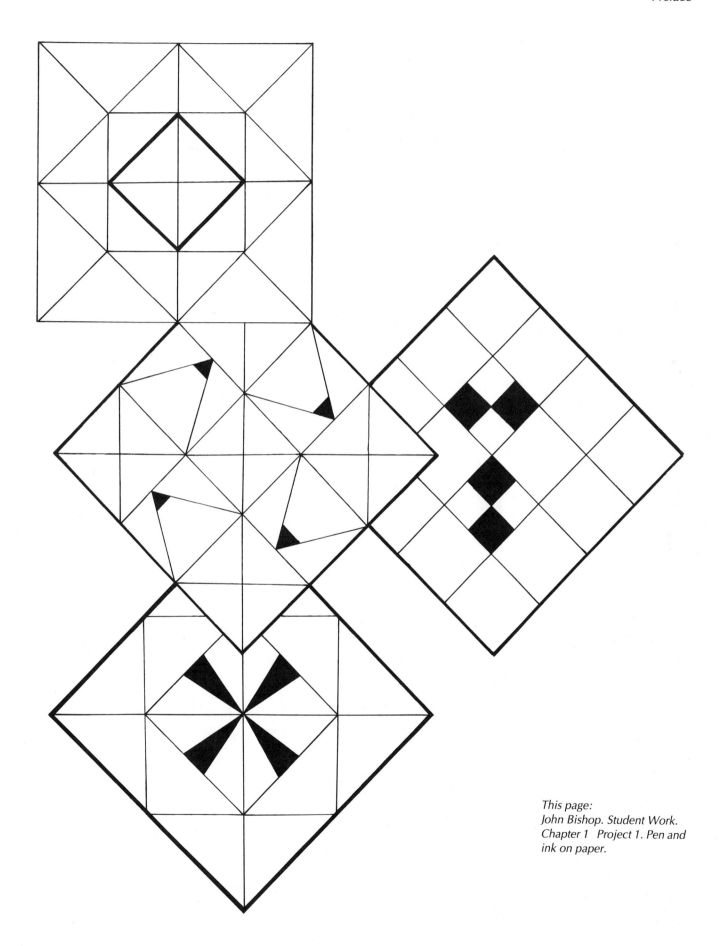

This page:
John Bishop. Student Work.
Chapter 1   Project 1. Pen and
ink on paper.

Introduction

Nature is relationships in space.
Geometry defines relationships in space.
Art creates relationships in space.

This book is about space, the arena of all life's interactions. It is the common denominator of Mathematics, Art and Nature. Space, however, cannot be perceived without our initially forming a space conception; an idea that first defines what it is. The experience of physical space exists for everyone, but the idea is an abstraction; a way to step outside the experience and then attempt to define it through words and symbols.

To the mathematician the term space is chosen to be undefined, as are point, line and plane. Mathematicians are concerned with the development of logical symbolic structures, one of which is geometry. In playing the game of geometry, the undefined terms are considered to be the first game pieces. One of the rules of the game dictates that terms (new game pieces) may be added only if they can be defined by terms which are either undefined or have been previously defined. This rule prevents the game from becoming circular in its logical structure.

Another rule in this game allows the player to formulate statements which will be accepted as true without proof. These statements are called postulates and axioms. Yet another rule requires that the number of these be kept to a minimum and never be used when a theorem, which can be proved, may be used instead.

This game was originally played informally by the Classical Greek philosophers. It was Euclid, circa the third century B.C., who wrote *Elements*, the rule book for plane geometry which describes two-dimensional figures and relationships between them. New strategies were developed by later players who altered one of the postulates (the fifth to be precise) giving rise to new geometries that were no longer played on a flat board.

Mathematics is one tool by which we attempt to understand how the natural world operates. As our comprehension of this world increases, the mathematical models need to change in order to adapt to that new understanding. Ancient geometry originated from the practical need to survey and measure the land which was perceived as flat. Current models have expanded to fit the concept of a universe composed of curved surfaces, ever changing in relationship to each other.

Thinking about space requires the ability to reason abstractly. It is difficult since the "nothingness" of space is everywhere and nowhere is it visible. It is intangible and, yet, we must give it tangibility in order to examine it. Moreover, not all qualities of space are accessible to sense perception. We cannot comprehend large-scale or small-scale space without the aid of sophisticated equipment such as telescopes or microscopes.

Most of our experience of the external world comes through to us via our senses, with vision and touch playing the most significant roles. When we speak of space we must consider the forms that are placed within it. How humans situate objects in relation to each other is instinctive and operates at a subconscious level. Art objects, on the other hand, are symbolic representations of spatial experiences. Various periods throughout the history of art show us changing concepts. Humankind's views of space go beyond the physical into the realms of the emotional and the metaphysical as expressed through art objects. How we humans

envision our position in the cosmos can be comprehended through the graphic, sculptural and architectural projections of a particular cultural period.

Out there in the cosmos, space is continuous and multidirectional. All directions are equally valid and all reference points equally important. There is no reference to a "me." In here, we live in a three-dimensional world. It has length and width and depth. We continually move through our space and our movement plays a decisive role in our perception of it. By touching and seeing, form becomes accessible. With our ever roving eyes,

our ever moving bodies, our ever active limbs, we experience a dynamic, not static, environment, which is space.

As physical creatures in this three-dimensional space, our understanding is tempered by both our physical and perceptual selves. We no longer walk on all fours. Because of this upright posture, verticality has become the dominant direction along with its complement, the horizontal. A person with an outstretched arm, standing on the ground plane, could allow for the development of the abstract idea of the "right angle". A person pivoting on one leg could lead the way for the development of the concept of "circle." Notions of measure and distance in space grow from our human scale and the understanding of our physical place in this space. Each person is a point of orientation and the center of the universe.

In Art, the vertical became the universal organizing principle that has reigned supreme since the Neolithic period up to, and including, the present. This vertical has come to stand for the line of movement and of striving upward while, in contrast, the horizontal suggests repose and balance. But we must remember these are relative positions. There are other ways to portray order.

Our two eyes, located in the fronts of our heads, apprehend position in space. We move our heads right and left giving us only 180 degrees of vision before we must turn around to deal with back and front. Thus, our limited frame of movement forces us to define space in a particular fashion. Gravity pulls upon our bodies so that the ideas of up and down are determined by that force. Here and there are notions of location. Near and far involve the concepts of time and movement through space.

Our brains and perceptual processes evolved to suit our own particular brand of space. This medium became our message. We did not learn to really "see" our space until the non-Euclidean mathematicians pointed out that there were other spaces whose forms and patterns differ from ours in very significant ways. Mathematically, we can have spaces with more than three dimensions which allow for the examination of certain problems that could not be solved within the structure of Euclidean geometry. At this time, however, the artist-designer works within a relatively limited spatial range. Two and three-dimensional spaces are the most practical for generating design structures.

All of Nature's interactions occur in a spatial field, and there are structural

Every thing made by human hands looks terrible under magnification—crude, rough, and unsymmetrical. But in nature every bit of life is lovely. And the more magnification we use, the more details are brought out, perfectly formed, like endless sets of boxes within boxes.

Dr. Roman Vishniac

We see that certain patterns bring certain benefits and efficiencies, irrespective of the size of the system, the forces, or the particular materials used.

Peter Stevens

This page:
Natural patterns

Top to bottom:

Branching
   Short and relatively direct, it fills space. It relieves congestion at the center while providing efficient paths from the center to outlying points.

Cracking
   The pattern that is created by tension on a surface.

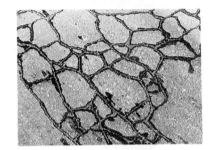

Explosion
   The radiating shape that occurs wherever numerous lines fan outward from a single, central point. It minimizes the distance between the center and the outlying points and provides increased surface area.

Meander
   A line that bends and curves between two fixed points. Structurally, a meander provides strength.

Spiral
   It uniformly fills space in a regular manner as it curves around itself. It allows a maximum amount of material to fit within a fixed space.

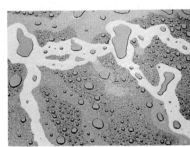

constraints that allow for only a limited number of basic patterns. They occur as a result of Nature's practicality since everything she does seems to have a purpose. She works within the interaction of many complex forces where conditions are never simple. Spirals, meanders, explosions, branchings, and crackings solve different spatial problems with the greatest efficiency. These patterns recur with infinite variation at the microscopic and macrocosmic levels, but they swerve from the absolutely perfect geometric model which, after all, is only an abstraction created by the human mind in order to parallel experience.

In creating forms in space, Nature does not indulge in sloppy craftsmanship. All details of design are fully worked out with a fusion of function, technique and material. She gives the same amount of attention to small and large, organic and inorganic forms alike, and no aspect of her design is unimportant. Her packing and partitioning of space are ultimately practical and efficient. Beauty is the by-product of her design as a result of the form and its function being intimately related.

Whatever we may say is obviously inadequate to the reality of Nature, for, as Jacob Bronowski has said, "The world is totally connected. Whatever explanation we invent at any moment is a partial connection, and its richness derives from the richness of such connections as we are able to make."

Through this book, we hope to encourage a reverence for all aspects of Nature's being.

Further Reading

Bronowski, Jacob. *The Origins of Knowledge and Imagination*. New Haven: Yale University Press, 1978.

Bronowski, Jacob. *Science and Human Values*. New York: Harper and Row, 1965.

Capra, Fritjof. *The Tao of Physics*. New York: Bantam Books, 1980.

Capra, Fritjof. *The Turning Point*. New York: Simon and Schuster, 1982.

Feininger, Andreas. *The Anatomy of Nature*. New York: Dover Publications, Inc., 1979.

Feininger, Andreas. *Nature Close-Up: A Fantastic Journey Into Reality*. New York: Dover Publications, Inc., 1981.

Jammer, Max. *The History of Theories of Space in Physics*. 2nd Ed. Cambridge: Harvard University Press, 1969.

Judson, Horace Freeland. *The Search for Solutions*. New York: Holt, Rinehart and Winston, 1980.

Kline, Morris. *Mathematics in Western Culture*. New York: Oxford University Press, 1953.

1 Basic Constructions

> *Creation directly from principles, and not through the imitation of appearances, is the real way to freedom for an artist. Originality is the product of knowledge, not guesswork.*
>
> Joseph Schillinger

Mathematics flourishes in an environment where there is unlimited freedom of inquiry regardless of whether the results have immediate practical application. Such was the intellectual climate of classical Greece. The culture was that of a fabric tightly interwoven with the strands of philosophy, mathematics and art. Each enhanced and clarified the other as well as defined the position of humans in their relationship to Nature. Unlike all preceding civilizations, the Greeks were the first to claim that Nature was comprehensible to humans through the use of reason and through the application of mathematics.

Though the elements of geometry were known by the Egyptians and Babylonians, and transmitted to the Greeks, it was the latter who created a mathematics that far surpassed that of their predecessors and which has lasted to this day. For beings endowed with a reasoning faculty, there was a method of gaining knowledge that was different from induction based on experience. Deduction allowed one to draw conclusions based on a few premises.

In spite of limits to this method of reasoning, the Greeks insisted that all mathematical conclusions be obtained by the deductive process alone. They were naturally drawn to this approach by their involvement with the mental activities of philosophy which dealt with abstract questions and broad generalizations. The plane geometry of Euclid with its ten axioms is the supreme example of this method. The many theorems of this book grew from a few original statements.

The love of rationality led to the Greek concept of beauty in the visual arts. In sculpture and architecture they sought the ideal, the eternal, the perfect, the complete, the consistent, the pure. What they sacrificed was the individual, the ephemeral, the changing, the imperfect. Theirs was a static world of closed form. It was reason, and not unreliable senses, that was to govern their way of thinking. With the use of reason, mathematics became involved with the abstract, whereby mental concepts rather than physical things were essential. A theorem could be used in a great many cases, removing its connection to a specific concrete thing and applying it to an ideal form. Theorems were considered permanent, perfect and incorruptible, unlike the experiences of the senses. Today we understand that our senses give us access to a limited range of experience only. It is through reason and deduction that scientists have been able to discover the electrical, atomic, and subatomic aspects of our universe, all of which reside in realms not accessible to our senses.

When the Greeks realized that the harmonies of the Universe could not always be explained in terms of whole number ratios, as they had originally thought, plane and solid geometry became the subjects of mathematical emphasis. They were investigated because figures and shapes had constant properties. Instead of dealing with numbers, everything was conceived of in terms of geometry, whereby a length was designated as the

Below:
Sharon Maniates. Student Work. Chapter 1 Project 1. Pen and ink on paper.

number *1* and then other numbers could be represented in terms relating to this length. Thus, "answers" could be expressed in terms of figures rather than numbers. It was left to other times and cultures to extract the essences of numerical relationships.

Need arose, then, for a method of accurately constructing the figures under examination. Liking a challenge, the Greeks set very tight limits on which tools were permissible for construction. The essential ones were the compass and straightedge and, with a few notable exceptions, almost all of the figures that were dealt with could be constructed using these.

Interestingly, there is speculation that the limitations on construction tools came about because the Greeks, preferring mental activities, shunned working with their hands, an occupation for women, slaves and artisans. Occasionally they extended the challenge, setting further restrictions on their tools, such as using a compass that only had one setting; the so-called *rusty compass.* Without going to this extreme, we invite the reader to experience the challenge of the Greeks while examining the harmonious relationship that exists between the worlds of Art, Mathematics and Nature.

In order to better understand the concepts explored in this book, hands-on experience is essential. At the end of each chapter we have provided problems and projects. Each enables you to physically engage the ideas presented. The problems pertain to the mathematical ideas and usually require less time to complete. The projects are designed to join the mathematical ideas to the

forming abilities of the artist. They are meant to give aesthetic satisfaction as well as to present a challenge.

Because of its precision, we have chosen to use mathematical language. Since lay persons and mathematicians frequently give different meanings to a term, we will point out the discrepancies in interpretation whenever possible.

In the back of the book we have provided a glossary, which not only defines terms mathematically but also gives diagrams and symbols for them. Acquaint yourself with it because terms and symbols will frequently be used in the body of the text without definition. Appendix A explains the meaning of symbols. Appendix B provides further explanation of some of the mathematical concepts presented. Some helpful formulas are given in Appendix C, while Appendix D gives information on art materials and techniques.

Answers to selected problems can be found in Appendix E. We are firmly convinced that intuition, unlike instinct, is rationality at a deeper level. Therefore, this appendix is designed to facilitate the "cementing" of the mathematical ideas so that they become part of your internal vocabulary. Greater detail is given in solutions to problems in the first chapters, and lessens with successive ones. Practice will enhance comfort, competency, and skills while securing the concepts. Try the problems before turning to the answer section, and remember that most may well have more than one solution.

Since we are investigating the properties of physical space and the ways in which we may apply Nature's laws to gain aesthetic divisions of that space, a knowledge of certain basic geometric construction techniques is necessary. Using a T-square and a draftsman's triangle frequently simplifies construction problems, since, with these tools, perpendicular and parallel lines are quite easily drawn. However, many constructions cannot be handled with a T-square and a triangle. These frequently require knowledge of basic compass and

Below:
Kelly Ross. Student work. Chapter 1 Project 5. Pen and ink on paper.

straightedge techniques for their execution. We have, therefore, presented all constructions as compass and straightedge problems.

Quality tools directly influence the quality of the finished product. We suggest that you use a good compass; a metal straightedge that is elevated to prevent smudging; an HB pencil with a good sharp point or a mechanical pencil with HB lead; and a Pink Pearl eraser. Heavy weight bond paper is an appropriate type to use for the problems, but projects may require heavier paper. Other tools and supplies will be suggested as needed.

Before you start, make sure the tips of your compass are aligned and that you have a smooth surface to work on. Keep in mind as you work that larger constructions are usually easier to do and are generally more accurate.

In the section that follows, eight basic constructions are presented with points labeled as an aid in understanding the steps. Labels are not an essential part of constructions, but you may want to use them at first to gain insight into the techniques involved. Later on, when using constructions as a basis for a project design, you will find labelling unnecessary.

The only permissible tools for the following constructions are: *compass, straightedge* (not a ruler which has markings for specific measurements), *pencil* and *eraser*. You will notice a similarity in format. When you see the word *Given,* it indicates the figure with which you have to start. *Given* figures in the diagrams, which are visually bolder, have specific measures. When you practice the constructions, however, the *given* figure that you draw may well have a different measure, eg. a shorter line segment, a larger angle, a shift in position, etc. To help you understand the procedure, extra information is provided in these first constructions, and the diagrams give step-by-step information. As you become more comfortable with the process, less additional information is added.

Right:
Construction tools: straight-edge, pencil, and compass. Don't forget an eraser.

A ————————————————————— B

This is the given. In this and all future constructions the given figure will be drawn in a heavier line width.

C •————————————————————— D

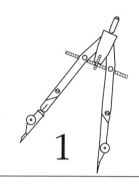

1

Construct a Line Segment Congruent to a Given Line Segment

Given line segment AB ($\overline{AB}$)

1. Begin by drawing what is given. The figure you draw may be different from the one shown in the text. You will work from the figure you draw.

All words in bold face type can be found in the glossary and all mathematical symbols are explained in Appendix A.

2. Now draw a **line segment*** anywhere in the plane. This should be longer than the given, $\overline{AB}$.

3. On this segment, choose a point C to be designated as one endpoint.

4. Place the metal tip of the compass on A and the pencil tip on B respectively to measure the length of $\overline{AB}$.

5. Without changing the compass setting, place the metal tip on C and cut an **arc** with the pencil tip that intersects the segment. Label the point of intersection D.

Now segment CD is **congruent** to segment AB. ($\overline{CD} \cong \overline{AB}$).

Note: The compass is used for measuring in Euclidean constructions. Equal lengths are obtained when the setting is left unchanged. Position in the plane has nothing to do with the measure of a figure.

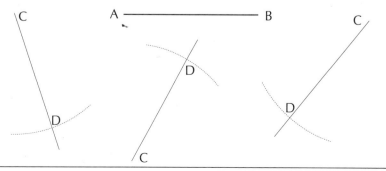

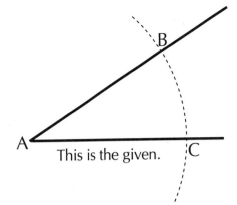

This is the given.

Given angle A (∠A)

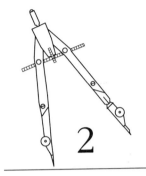

2

Construct an
Angle
Congruent to a
Given Angle

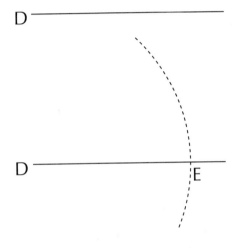

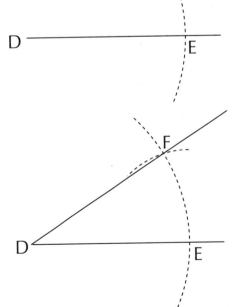

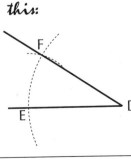

*Had the other
endpoint been
chosen to be D,
the construction
would look like
this:*

1. After drawing the given angle, you will construct one just like it. To do this you must first draw a line segment anywhere in the plane. Label one endpoint D.*

2. Put the metal tip of the compass on A and cut an arc with the pencil tip that intersects both sides of ∠A. Label these points of intersection B and C.

3. Without changing the compass setting, place the metal tip on D and cut an arc with the pencil tip so that it intersects the line segment. Label this point of intersection E.

4. Put the metal tip of the compass on C and adjust the setting so that the pencil tip rests on B. Without changing the setting, put the metal tip on E, and cut an arc that intersects the arc through E. Label this point of intersection F.

5. Draw **ray** DF ($\overrightarrow{DF}$).

Now angle D is congruent to angle A (∠D ≅ ∠A).

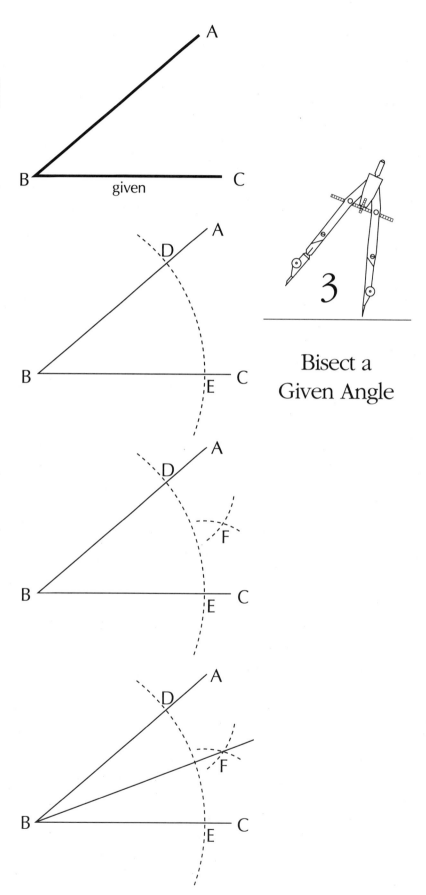

Given angle ABC (∠ABC)

1. Draw the given angle anywhere in the plane, and label it ABC.

2. Place the metal tip of the compass on B and cut an arc that intersects both sides of ∠ABC. Label these points of intersection D and E.

3
Bisect a Given Angle

3. Open the compass to more than half the distance from D to E. Place the metal tip on D and cut an arc in the **interior** of ∠ABC.

4. Without changing the compass setting, place the metal tip on E and cut an arc that intersects the one drawn in the previous step. Label the point of intersection of the two arcs F.

5. Draw ray BF (⃗BF).

Now angle ABF is congruent to angle FBC (∠ABF ≅ ∠FBC).

⃗BF **bisects** ∠ ABC and is called the **bisector** of ∠ABC.

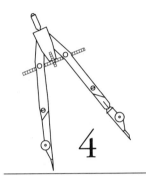

4

Construct a Perpendicular to a Line Through a Point on the Line

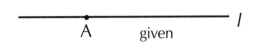

given

Given point A on line *l*

1. Place the metal tip of the compass on A and cut arcs that intersect *l* on both sides of A. Label the points of intersection B and C.

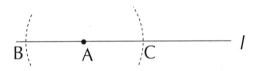

2. Open the compass to more than half the distance from B to C. Place the metal tip on B and cut an arc on one side of line *l*.

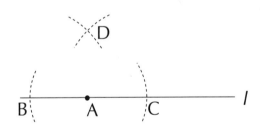

3. Without changing the compass setting, place the metal tip on C and cut an arc that intersects the one drawn in the previous step. Label the point of intersection D.

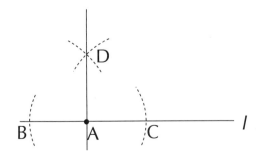

4. Draw line AD ($\overleftrightarrow{AD}$).

Now line AD is **perpendicular** to line *l* ($\overleftrightarrow{AD} \perp l$).

• A

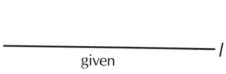

Given line *l* and point A *not* on *l*

given

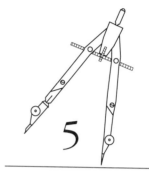

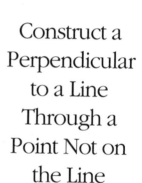

5

• A

1. Open the compass to more than the distance from A to *l*. With the metal tip on A cut two arcs that intersect *l*. Label the points of intersection B and C.

Construct a Perpendicular to a Line Through a Point Not on the Line

2. Place the metal tip on B and cut an arc on the side of *l* opposite A.

•A

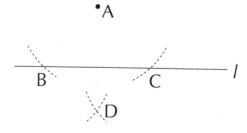

3. Without changing the setting, place the metal tip on C and cut an arc that intersects the one drawn in the previous step. Label the point of intersection D.

4. Draw line AD (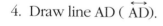).

Now line AD is perpendicular to line *l* (AD ⊥ *l*).

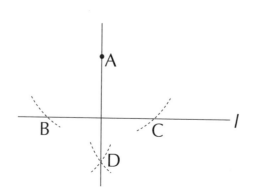

These facing pages:
Linda Maddox. Student work.
Technical pen and ink. These
hand drawings illustrate designs based on Constructions
1-5.

A •

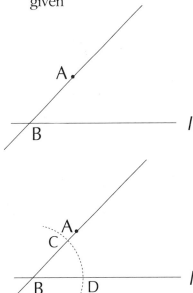

_____ *l*

given

| Given line *l* and point A *not* on *l* |

6

Construct a
Line Parallel to
a Given Line
Through a
Given Point

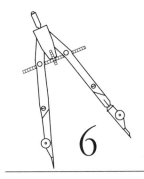

*Note: position
in the plane
and the angle
formed by l and
line AB does not
affect the con-
struction.*

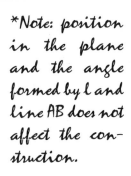

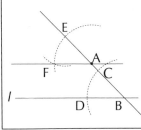

1. Draw a line through A that intersects *l.** Label the point of intersection B.

2. Place the metal tip of the compass
on B and cut an arc that intersects
both $\overleftrightarrow{AB}$ and *l.* Label the points of
intersection C and D.

3. Without changing the compass setting, place the metal tip on A and cut
an arc using the same orientation as
the arc drawn in the previous step
Label the point of intersection E.

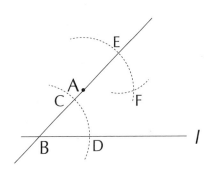

4. Place the metal tip of the compass
on C and the pencil tip on D. Without
changing the setting, place the metal
tip on E and cut an arc that intersects
the arc previously drawn. Label the
point of intersection F.

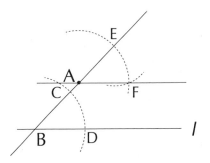

5. Draw line AF ($\overleftrightarrow{AF}$).

Now line AF ($\overleftrightarrow{AF}$) is **parallel** to line *l.*

Given line segment AB (\overline{AB})

A ———————————————— B
　　　　given

1. Open the compass to more than half the distance from A to B. Place the metal tip on A and cut arcs above and below $\overline{AB}$.

A ———————————— B

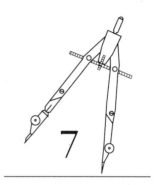

7

2.   Without changing the compass setting, place the metal tip on B and cut arcs above and below $\overline{AB}$ so that they intersect the other two arcs. Label these points of intersection C and D.

×C

Bisect a Given Line Segment

A ———————————— B

3. Draw line CD ($\overleftrightarrow{CD}$). Label the point of intersection E.

Now line segment AE is congruent to line segment EB ($\overline{AE} \cong \overline{EB}$).

×D

$\overleftrightarrow{CD}$ **bisects** $\overline{AB}$ and is called the **bisector** of $\overline{AB}$.
E is the **midpoint** of $\overline{AB}$.
Also, $\overleftrightarrow{CD} \perp \overline{AB}$ and, therefore, $\overleftrightarrow{CD}$ is the **perpendicular bisector** of $\overline{AB}$.

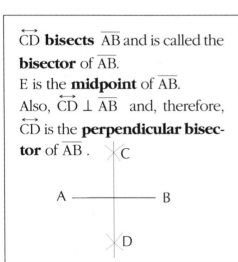

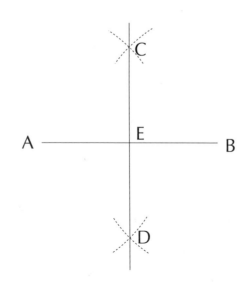

This page:
These photographs of architectural details show several of the basic constructions used in a design format. You should be able to find parallel and perpendicular lines, congruent segments and angles, and perpendicular bisectors.

Frequently it is useful to be able to divide a line segment into any number of congruent segments. If the number of segments desired is a power of two, i.e., 2, 4, 8, 16, 32, etc., it is easiest to bisect the segments and continue bisecting the newly formed segments.

| Given line segment AB ($\overline{AB}$) |

A ———————————————— B

given

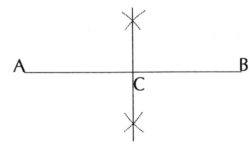

1. Using the information from Construction 7, bisect $\overline{AB}$. Label the point of intersection C.

8

Divide a Line Segment into Congruent Segments Numbering Powers of Two

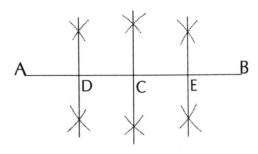

2. Bisect $\overline{AC}$ and $\overline{CB}$. Label the points of intersection D and E.

Now $\overline{AD} \cong \overline{DC} \cong \overline{CE} \cong \overline{EB}$.

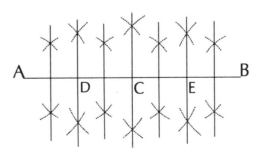

Continue to bisect line segments until you have obtained the desired number of congruent segments.

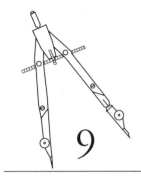

Divide a Line Segment Into a Given Number of Congruent Segments

A —————————— B   Given line segment AB ($\overline{AB}$)
given

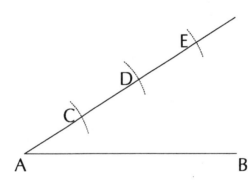

1. Draw a ray from A to form an **acute*** angle with $\overrightarrow{AB}$.

2. Mark off a given number of congruent segments on the ray by first placing the metal tip on A and cutting an arc on the ray. Repeat without changing the compass setting, each time placing the metal tip on the point of intersection of the most recently cut arc and the ray. Label the points of intersection C, D and E, respectively.

(We have chosen to divide the line segment into three congruent segments for the purpose of demonstration. Any other number may be used.)

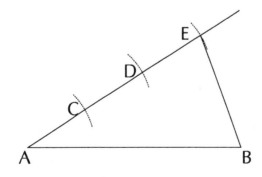

3. Draw $\overline{EB}$.

*The angle does not have to be acute, but the construction is easier to keep accurate if it is. Keep the lead point of your compass finely chiseled; a dull point produces inaccuracies!

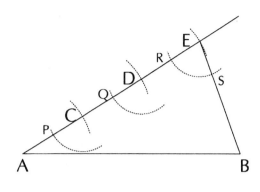

4. Place the metal tip on E and cut an arc that intersects both $\overline{EA}$ and $\overline{EB}$. Without changing the setting, cut arcs at D and C in the same orientation. Label points of intersection P, Q, R and S, as shown.

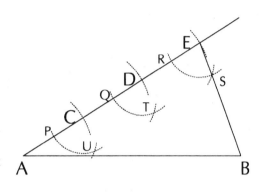

5. Place the metal tip on R and the pencil on S. Without changing the setting, put the metal tip on Q and cut an arc that intersects the one through Q. Repeat at P. Label points of intersection T and U, as shown.

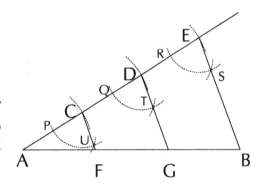

6. Draw segments through D and T and through C and U and extend to $\overline{AB}$. Label points of intersection F and G.

Now $\overline{AF} \cong \overline{FG} \cong \overline{GB}$.

Note: Once $\overline{CB}$ is determined, this measurement could be used to mark off the remaining congruent segments on $\overline{AB}$. This makes it necessary to copy only one angle. However, any inaccuracies in the length of $\overline{CB}$ will be magnified by the time you reach A, which can be very frustrating.

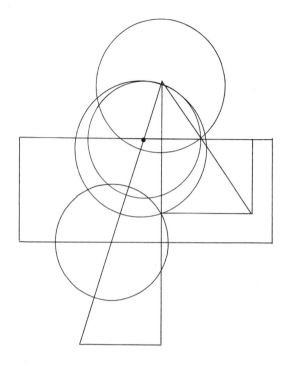

This page:
Linda Maddox. Student work.
Art images developed from
Constructions 6, 7 and 9.

Problems

1 Construct a triangle with sides congruent to segments a, b, and c respectively. *(Hint: Refer to Construction 1 for congruent segments and to Appendix B for conditions for congruent angles.)*

2 Construct a triangle congruent to △ABC. *(Hint: refer to previous problem.)*

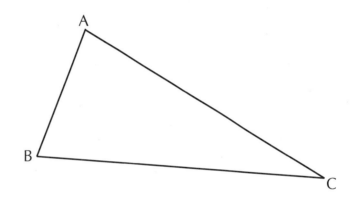

3 a. Construct a line segment congruent to $\overline{RS}$ and divide it into eight congruent segments. *(Hint: Refer to Construction 8)*

   b. Construct a segment congruent to $\overline{RS}$ and divide it into seven congruent segments. *(Hint: Refer to Construction 9)*

4 Construct all three **altitudes** (see Glossary) of △ABC. *(Hint: Refer to Construction 5)*

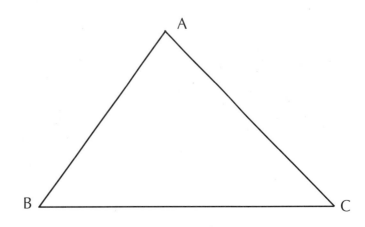

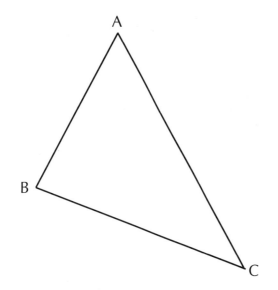

5 Construct a **rectangle** with width equal to MN and length equal to OP. *(Hint: Refer to Constructions 4 and 1.)*

6 Construct the angle bisector of each of two angles of ∆ABC. Label the point of intersection P. Construct a perpendicular from P to $\overline{AB}$. Label Q, the point of intersection on $\overline{AB}$. Using $\overline{PQ}$ as a **radius**, draw circle P. *(Hint: Refer to Constructions 3 and 5.)*

This page:
James Carson. Student Work. Cut paper.

Opposite page:
Ann MacLean. Student work. Colored pencil on colored paper.

"Round Robin" constructions. In this project each student in the class was instructed to execute a particular construction and then pass his or her paper to the right. The process was repeated several times. Students were then instructed to take the papers and embellish them so they became more artful.

7 Construct the perpendicular bisector of each of two sides of ΔABC. Label the point of intersection O. Using $\overline{OA}$ as a radius, draw circle O. *(Hint: Refer to Construction 7)*

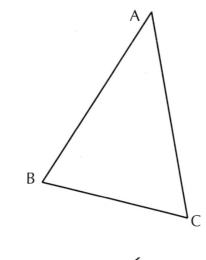

8 Construct two different sized triangles each of which has two angles congruent to the two given angles (A and B). *(Hint: Refer to Construction 2)*

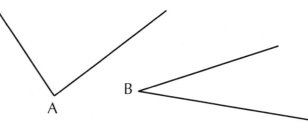

9 On $\overleftrightarrow{AB}$ construct lines parallel to $\overleftrightarrow{CD}$ through points A and B respectively. *(Hint: Refer to Construction 6)*

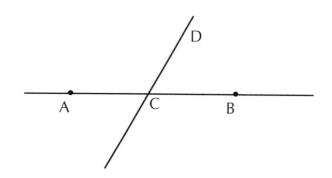

Projects

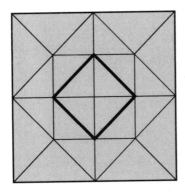 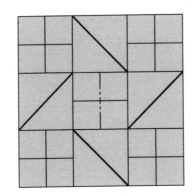 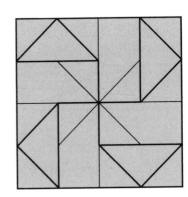

1 Choose one of the geometric designs above and duplicate it in a square *(Check Appendix B for the properties of a square.)* measuring five inches on a side. Repeat the construction a minimum of *three* times making variations on the original each time. Put the four (or more) blocks in finished form by doing them in black and white or color and by being concerned with the arrangement.

2 On heavy paper or illustration board, construct a rectangle that measures 8" x 13" *(Check Appendix B for the properties of a rectangle)*. Create a design with the following restrictions:

a. The design must be done within the limits of this constructed rectangle.

b. It must use at least *three* of the basic constructions you have just mastered.

c. It must be done in black and white only.

3 Using the instructions for Project 2 as a basis, create a solution using color.

4 On heavy paper, or illustration board, construct a square that measures 13" x 13". Then, with compass and straightedge, construct a figure similar to one of those on the next page. Give the figure color by using materials of your choice (fabric, colored pencils, paints, inks, construction paper, etc.)

5 Using multiples of a variation on *one* of the squares on this page *(a photocopier would be useful here)*, create a paper "quilt" of three blocks by four blocks. It may be in black and white only or use a color scheme of your choice.

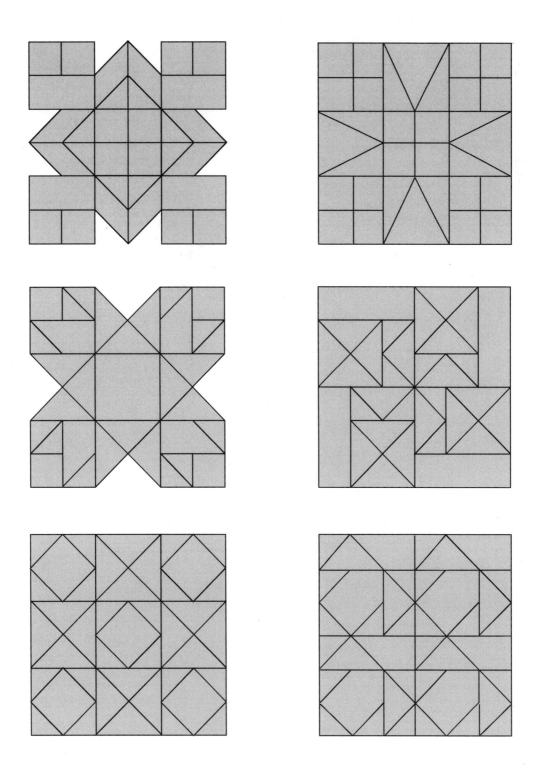

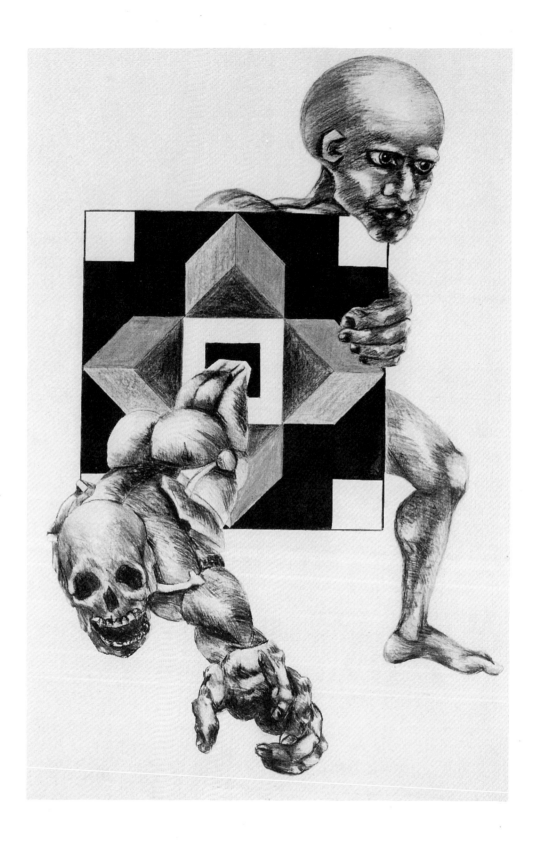

This page:
Sharon Summers. Student
Work. Chapter 1 Project 4.
Colored pencil on illustration
board.

Further Reading

Bruni, James V. *Experiencing Geometry.* Belmont, California: Wadsworth Publishing Company, Inc., 1977.

Farrington, Benjamin. *Greek Science.* Nottingham: Spokesman, 1980.

Gardner, Martin. *Geometric Constructions with a Compass and Straightedge, and Also with a Compass Alone. Scientific American.* Vol. 221, no.3, September, 1969.

Seymour, Dale and Reuben Schadler. *Creative Constructions.* Palo Alto, California: Creative Publications, 1974.

2 *Unique Relationships*

For the visual artist, a primary concern is for the division of space, be it two- or three-dimensional. How does one go about dividing a space so that there is harmony between the smaller elements and the whole? There must be some way to determine the essential relationships beyond intuition alone. The understanding of the concept of proportion leads to more rational solutions to the problem. Reason and intuition must function simultaneously.

One of the most meaningful avenues of investigation is found in examining proportions within the human body. Each age redefines the *body beautiful*; however, many of our ideas about correct proportions come down through the ages to us from Classical Greece. When working with the nude form the Greek artist was fundamentally interested in the ideal; articulating a concept of a perfect shape and adapting natural form to this idea. The famous Greek sculptor Polyclitos was concerned with the idea of completeness, balance and clarity, communicated through the nude form of the male athlete. It is believed that he wrote about a system of proportions in a book presumed lost.

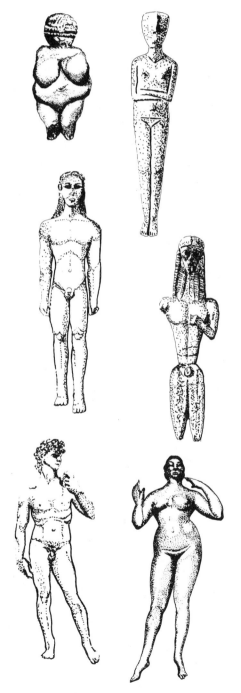

This page:
Fig. 2.1 Sculptures displaying various systems of proportions. Which works do you find beautiful?

Top left:
Venus of Willendorf. *15,000-10,000 B.C. Stone. 4.375".*

Top right:
Idol from Amorgos. *2500-1100 B.C. Marble. 30".*

Center Left:
Standing Youth. *Kouros, ca. 600 B.C. Marble. 6.5'.*

Center right:
Mantiklos Apollo. *7th century B.C. Bronze. 8".*

Bottom left:
Michelangelo. David. *1501-1504. Marble. 13' 5".*

Bottom right:
Gaston Lachaise. Standing Woman. *1912-1927. Bronze. 70".*

8 Heads   **7.5 Heads**

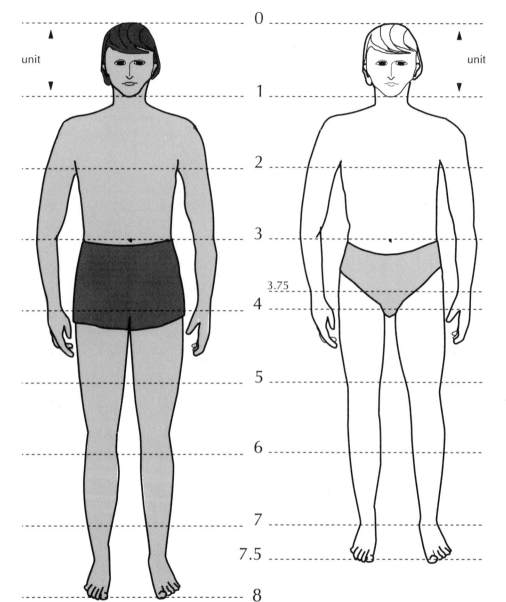

This page:
Fig. 2.2 Comparison of two systems of body measurements. In both, the head is used as the basic unit.

Left:
For Praxiteles, the Hellenistic Greek sculptor, ca. 340 B.C., and the Italian Renaissance artist Michelangelo, 16th Century A.D., the ideal was 8 heads.

Right:
Phidias, an earlier Greek sculptor, determined that the ideal body was based on a system using a total of 7.5 heads. His concept is still used by many artists today with:

a. the waist determined at 3 heads
b. the hips at the halfway point of 3.75 heads
c. the legs being considered one-half the length of the entire body
d. the knees at 5 heads
e. the ankle at 7 heads
f. ankle to foot at one-half head.

What other systems could you develop?

What does proportion mean?

This is a term we use frequently in our everyday lives; one that brings to mind the qualities of harmony, balance, and fitness. We sense that something is "out of proportion" when it feels "wrong" and we tend to describe that condition as ugly.

The artist talks about proportions, suggesting the relationship of the parts to the whole, but never articulates a precise definition except as expressed through a given artwork.

Language tends to break down in discussing this idea. It is no wonder that the expression "beauty is in the eye of the beholder" has become a catchphrase or a last defense. It is assumed that an object is without intrinsic relationships and is only given meaning through the viewer.

Even dictionaries have unclear definitions:

1. *the comparative relation between things or magnitudes;*

2. ***proportions***, *dimensions or size;*

3. *any portion or part in its relation to the whole;*

4. *symmetry or balance.*

These definitions are not only circular but move away from specifics and become more and more general.

Through the course of our research we have discovered that Nature tends to rely repeatedly on particular relationships. Therefore, we are more strongly committed to the idea that Nature provides the model for beauty, and mathematics describes it. The dictionary gives us the feeling that proportion is a relationship between two things. Mathematics tells us precisely what that relationship is.

Before defining proportion, it is necessary to understand the term **ratio**. A ratio is the comparison of numbers to each other which can be expressed in several ways. If, for example, we are comparing the number 2 to the number 3, we may write 2 to 3, or 2:3, or simply 2/3. Hence, every fraction is a ratio in which the numerator is compared to the denominator. There are times when a ratio compares three or more numbers. Such is the case when discussing the relative lengths of the sides of a triangle. In such instances we use the notation a:b:c, where *a,b* and *c* represent the lengths of the sides. A ratio is only a quantitative comparison and not a qualitative one. The ratio of 3 elephants to 4 birds is precisely the same as the ratio of 3 golf balls to 4 galaxies, even though the things being compared are, themselves, very different.

Below:
Fig. 2.3 A ratio is a comparison of numbers, relating only to quantity. Here, the number 3 is compared to the number 4.

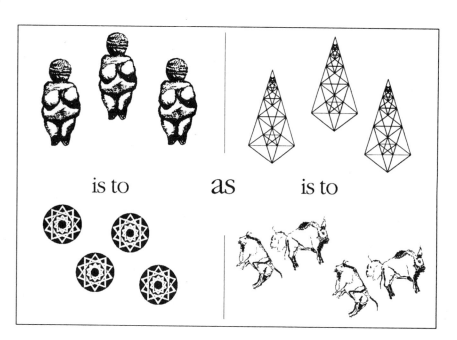

is to   as   is to

This page:
Fig. 2.4 The concept of the
"unit length" in relation to a
line segment.

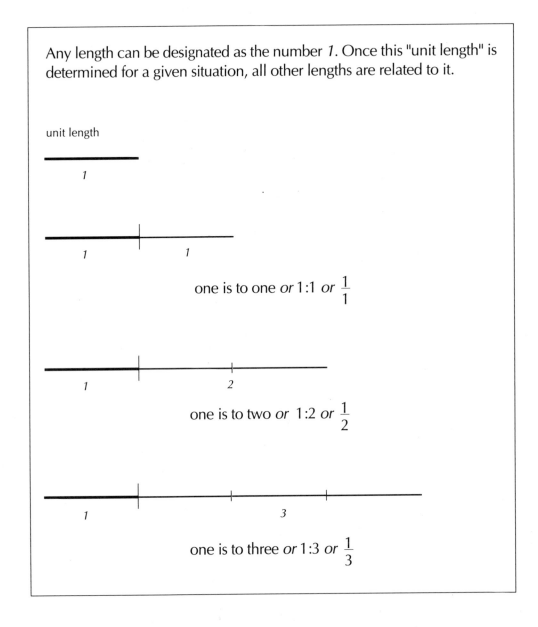

Any length can be designated as the number *1*. Once this "unit length" is determined for a given situation, all other lengths are related to it.

unit length

1

one is to one *or* 1:1 *or* $\frac{1}{1}$

1   2

one is to two *or* 1:2 *or* $\frac{1}{2}$

1   3

one is to three *or* 1:3 *or* $\frac{1}{3}$

A **proportion** is simply an equation in which two ratios are set equal to each other, such as 2/3 = 4/6 or 2:3::4:6 (read 2 is to 3 as 4 is to 6). When we write a/b = c/d , *a* and *d* are called the **extremes** of the proportion and *b* and *c* are called the **means**. The names of the terms become much more meaningful when we write a:b::c:d. The extremes, then, are the outside terms and the means are the middle terms. A **mean proportion** is one in which the means are equal, such as 1:3::3:9 or 1/3 = 3/9.

This page:
Fig. 2.5 The Divine Propor-
tion.

Top to Bottom:
Expressed mathematically.
Expressed visually.
Expressed with the longer seg-
ment at either end of the orig-
inal segment.

The Divine Proportion and the Golden Ratio In Relation to a Line Segment

One mean proportion which appears with amazing frequency in Nature and Mathematics, and has been used throughout the centuries by artists, is called the **Divine Proportion.** This proportion is derived from dividing a line segment into two segments with the special property that the ratio of the whole segment to the longer part is the same as the ratio of the longer part to the shorter part $(\frac{AB}{AC} = \frac{AC}{CB})$.

The ratio expressed by either side of the equation in Fig. 2.5 is called the **Golden Ratio** or the **Golden Mean.** The point which divides a line segment into two segments in the Golden Ratio is called the **Golden Section** or the **Golden Cut**, and is unique with the exception of order; that is, the longer segment may be at either end of the original segment.

The Golden Cut may be found geometrically by using Construction 10 which is found on the following page.

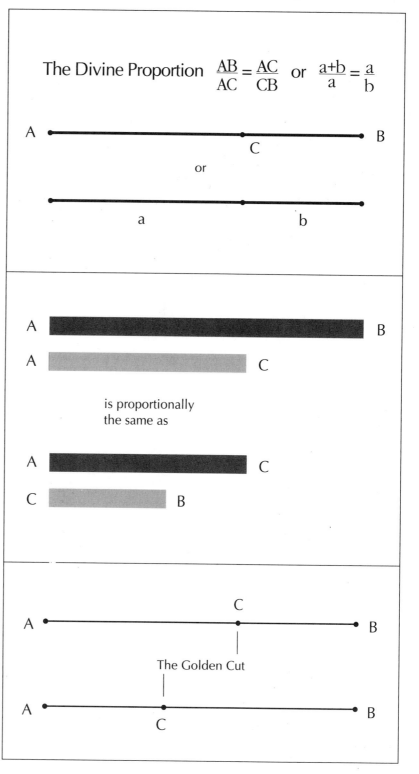

The Divine Proportion $\frac{AB}{AC} = \frac{AC}{CB}$ or $\frac{a+b}{a} = \frac{a}{b}$

or

is proportionally
the same as

The Golden Cut

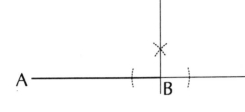

Given $\overline{AB}$

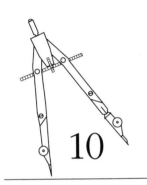

10

Construct the
Golden Cut
of a Line
Segment

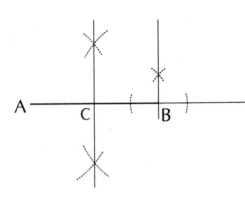

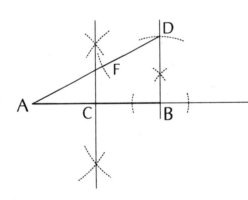

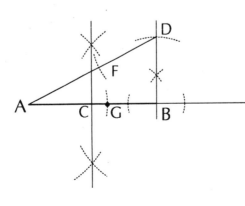

1. Extend $\overleftrightarrow{AB}$ through B.

2. Referring to Construction 4, construct a perpendicular to $\overleftrightarrow{AB}$ through B.

3. Referring to Construction 7, bisect $\overline{AB}$ and label the point of intersection C.

4. Use the compass to measure the distance between C and B. Without changing the compass setting, put the metal tip on B and cut an arc on the perpendicular constructed in Step 2. Label the point of intersection D.

5. Draw $\overline{AD}$.

6. Without changing the compass setting, put the metal tip on D and cut an arc that intersects $\overline{AD}$. Label the point of intersection F.

7. Put the metal tip on A and the pencil tip on F and cut an arc that intersects $\overline{AB}$. Label the point of intersection G.

Now G is the Golden Cut of $\overline{AB}$ and
$$\frac{AB}{AG} = \frac{AG}{GB}.$$

This page:
Student Work. Claire Melanson. Chapter 2 Project 1. Pencil on paper. Notice the use of the previous construction as the underlying structure for the image.

Below:
Linda Maddox. Student work. Chapter 1 Project 5. Three-dimensional female figure set inside a corner stage. Mixed media using technical pen and ink, recycled dungaree material, acrylic paint, machine and hand sewing.

In this text we have chosen to give equal emphasis to both mathematics and art. In order for an artist to use the Golden Ratio in an artwork, he or she needs only to understand the concept and the construction. However, students of mathematics will be interested in relationships and properties beyond the basics of definition and construction techniques. It is for this group that we have provided much of the following information. If you are among those who find mathematics difficult, do not be intimidated. As you continue to use them, the ideas, symbolic notation and techniques will become more comfortable.

The Golden Ratio as a Number

In the world, we see objects, events and people in relation to one another. All artworks are defined by their composition of relationships. Numbers are also expressions of relationships devoid of objects. The Golden Ratio can be expressed numerically as well as geometrically. But in order to find out what is the numerical designation of the Golden Ratio, you would have to solve the following equation:

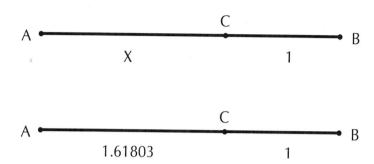

Let C be the Golden Cut of $\overline{AB}$.

We know then that $\frac{AC}{CB}$ is the Golden Ratio. We can find a numerical value for this ratio if we let CB = 1 and AC = x . Then $\frac{AC}{CB} = \frac{x}{1} = x$.

We can find x by using a little basic algebra.

From the Divine Proportion we know that $\frac{x+1}{x} = \frac{x}{1}$. Cross multiplying, we get the **quadratic equation** $x + 1 = x^2$ or $x^2 - x - 1 = 0$.

We can solve for x using the quadratic formula. (Check the glossary if the quadratic formula may have slipped your mind!)

$$x = \frac{1 \pm \sqrt{(-1)^2 - 4(1)(-1)}}{2(1)} = \frac{1 \pm \sqrt{1+4}}{2} = \frac{1 \pm \sqrt{5}}{2}$$

There are two solutions here: $\frac{1 + \sqrt{5}}{2}$ and $\frac{1 - \sqrt{5}}{2}$. However, $\frac{1 - \sqrt{5}}{2}$ leads to a negative number which does not have meaning when associated with the length of a line segment.

Therefore, we choose the positive solution, and $\frac{1 + \sqrt{5}}{2} = 1.61803$ correct to five decimal places.

> The Golden Ratio can be expressed approximately as:
>   1.61803 to 1
>   or 1.61803:1
>   or $\frac{1.61803}{1}$
> or simply 1.61803.

The number 1.61803 appears in the most surprising places and has such unique properties, mathematicians have given it a special name, **Ø**. This symbol is the Greek letter *Phi*, the first letter in the name Phidias, the Classical Greek sculptor who used the Golden Ratio so extensively in his work. This designation was chosen in his honor. Phidias was chief overseer of artistic enterprises under the Athenian leadership of Pericles in the Golden Age of Greece (5th century B.C.). He was considered responsible for the creation of the frieze panels, a continuous decorative band 525 feet in length, around the temple of the Parthenon, and was attributed with creating the huge ivory and gold statue of Athena, to whom the Parthenon was dedicated.

The Golden Ratio is Ø

$$\text{Ø} \approx 1.61803$$

Below:
Fig. 2.6 Shows a visual representation of the lengths of the segments resulting from the Golden Cut.

On any given line segment we can locate the Golden Cut. In order to be able to visualize the lengths of the resulting two segments, let us look at a unit line segment and compare 1.61803 to it.
1.61803 is greater than 1.5 (1.61803>1.5) but less than 1.75 (1.61803<1.75).

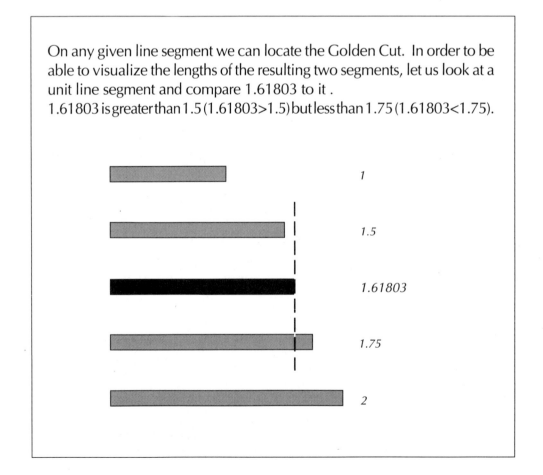

Top:
Rochelle Newman. Study of pentagons and pentagrams. Acrylic paint on illustration board.

Bottom:
Susan Hample. Student Work. Chapter 2 Project 2 variation. Metallic markers on paper.

Mathematical Properties of Phi

Since $\emptyset = \frac{1+\sqrt{5}}{2}$, $\frac{1}{\emptyset} = \frac{2}{1+\sqrt{5}}$ and $\frac{1}{\emptyset}$ is the reciprocal of $\emptyset$.

Correct to five decimal places, $\frac{1}{\emptyset} = .61803$.

Note that: $1.61803 - 1 = .61803$ or $\emptyset - 1 = \frac{1}{\emptyset}$.

Phi is the only number that becomes its own reciprocal when decreased by one. A proof of this property appears in Appendix A.

If you recall, when we solved the quadratic equation $x^2 - x - 1 = 0$, there were two solutions and we chose the positive one to represent $\emptyset$.

If we examine both roots we see that they bear an unusual relationship to each other.

As we know, $\emptyset = \frac{1+\sqrt{5}}{2}$.

Let us name the negative root $\emptyset'$ (phi prime) so $\emptyset' = \frac{1-\sqrt{5}}{2}$.

If we multiply the two together, we get:

$$\emptyset\emptyset' = \left(\frac{1+\sqrt{5}}{2}\right)\left(\frac{1-\sqrt{5}}{2}\right) = \frac{1-5}{4} = \frac{-4}{4} = -1.$$

That is, $\emptyset$ and $\emptyset'$ are **negative reciprocals** since their product is -1.

If we add them we have: $\emptyset + \emptyset' = \frac{1+\sqrt{5}}{2} + \frac{1-\sqrt{5}}{2} = \frac{2}{2} = 1$, and again the number 1 appears inextricably linked with $\emptyset$.

Let us consider the sequence whose terms are consecutive powers of $\emptyset$.

$$1, \emptyset, \emptyset^2, \emptyset^3, ..., \emptyset^n, ...$$

This is a **geometric sequence** since each term is gotten by multiplying the preceding term by the same number, namely $\emptyset$.

$$1\emptyset = \emptyset$$

$$\emptyset\emptyset = \emptyset^2$$

$$\emptyset^2\emptyset = \emptyset^3$$

$$\emptyset^n\emptyset = \emptyset^{n+1}$$

If we examine the terms more closely, we see that it is the case that each term is also the sum of the two preceding terms. Therefore, it is at the same time a **summation sequence.**

Since $\emptyset^2 - \emptyset - 1 = 0$, we have $\emptyset^2 = 1 + \emptyset$

or $\emptyset^2 = \emptyset^0 + \emptyset^1$.

Then $\emptyset^3 = \emptyset^2\emptyset = (1 + \emptyset)\emptyset = \emptyset + \emptyset^2 = \emptyset^1 + \emptyset^2$,

and $\emptyset^4 = \emptyset^3\emptyset = (\emptyset + \emptyset^2)\emptyset = \emptyset^2 + \emptyset^3$.

$\emptyset^5 = \emptyset^4\emptyset = (\emptyset^2 + \emptyset^3)\emptyset = \emptyset^3 + \emptyset^4$.

This process could be carried on indefinitely (with no small amount of tedium involved!) and the result would continue to hold; that is,

$$\emptyset^n = \emptyset^{n-2} + \emptyset^{n-1} \quad \text{or} \quad \emptyset^n + \emptyset^{n+1} = \emptyset^{n+2}$$

To diminish masochistic impulses for the hard-to-convince, a proof is supplied in Appendix B, along with proofs of the other three properties.

In Chapter 5 we will be examining in detail the Fibonacci Sequence, the simplest of the summation sequences. We will look at the connection between $\emptyset$ and this very special sequence at that time.

Properties of $\emptyset$

$$\emptyset - 1 = \frac{1}{\emptyset}$$

$$\emptyset\emptyset' = -1$$

$$\emptyset + \emptyset' = 1$$

$$\emptyset^n + \emptyset^{n+1} = \emptyset^{n+2}$$

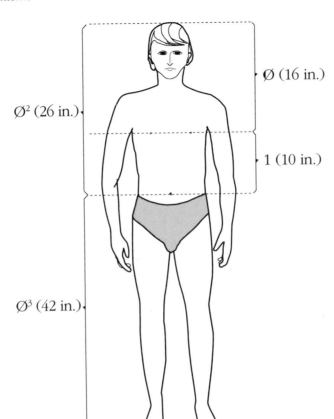

Ø (16 in.)

Ø² (26 in.)

1 (10 in.)

Ø³ (42 in.)

With this understanding of Ø, let us once again look at the human body as it conforms to Phidias' concept of the ideal. If we take ten inches as our unit of measurement and apply it to a man who is 68 inches tall, we find several lengths in Ø proportion. From the top of his head to the line of his breasts is 16 inches or Ø. (If the unit is 10 then 16 becomes Ø to one decimal place or 1.6). From his breasts to his navel is 10 inches or one unit. The measurement from his navel to the ground is 42 inches or Ø³. In the other direction, the distance between his navel and the top of his head is 26 inches or Ø².

Top:
Fig. 2.7 The standing figure illustrates Ø proportions in the human body.

Bottom:
Fig. 2.8 The face, too, can assume harmonic proportions.

$$\frac{AB}{CB} = \frac{AD}{ED} = \frac{DB}{DG} = \frac{DB}{FB} =$$
Ø ≈ 1.61803

$$\frac{CB}{HB} = \frac{CB}{IC} = Ø ≈ 1.61803$$

AD is half of AB.

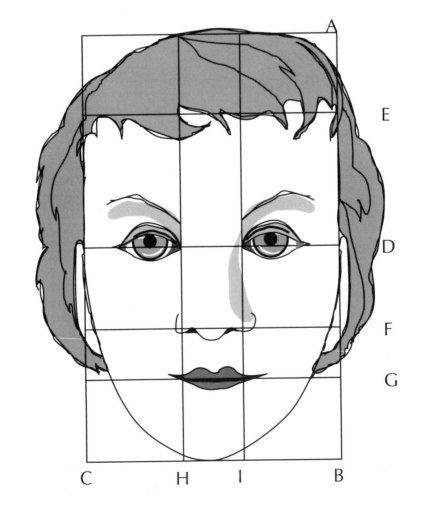

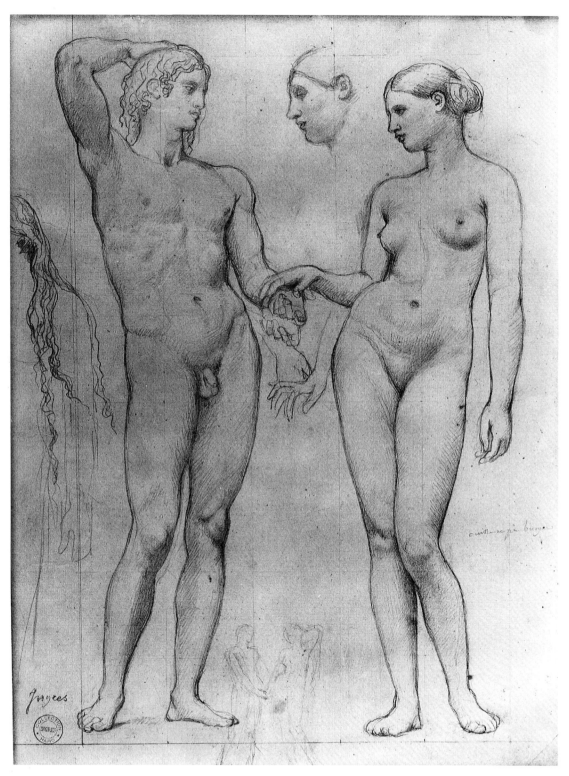

This page:
Jean Auguste Dominique Ingres. Two Nudes, *study for* The Golden Age, *1842. Graphite on paper, 16.375 x 12.375 ins. (41.6 x 31.4 cm). Courtesy of The Fogg Art Museum, Harvard University, Cambridge, Massachusetts. Bequest of Grenville L. Winthrop.*

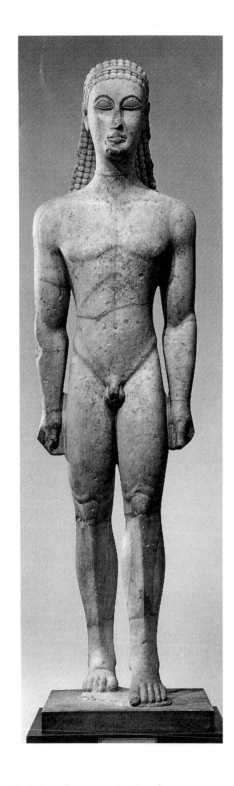
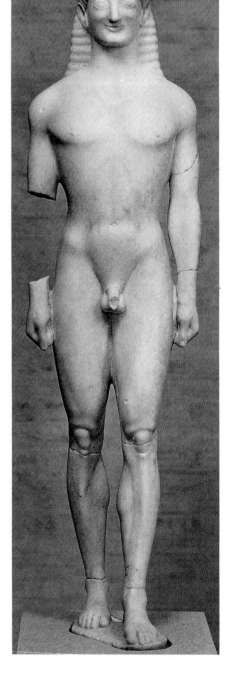

Facing pages:
Archaic and Classical Greek sculptures of the male figure.

Left to Right:
Kouros, c. 615-590 B.C. The Metropolitan Museum of Art, Fletcher Fund, 1932. (32.11.1)

Kouros from Tenea, c. 575-550 B.C. Staatliche Antikensammlungen und Glyptothek Munchen, Kat.- Nr. 168. (Foto Chr. Koppermann)

Kouros from Anavysos, c. 540-515 B.C. Athens, National Museum.

Doryphoros (the Spearbearer). Roman copy after an original by Polyclitos. c. 450-440 B.C. Marble, 6' 6". National Museum, Naples. Alinari/Art Resource, N.Y.

Subtle changes in body proportions are evident in Greek sculptures of the male figure from archaic to classical periods, despite the fact that the pose in all four figures is similar.

There seems to be a desire on the part of the artists to keep refining the proportions until the "ideal" has been met. The Classical norm appeared to indicate that the navel was the Golden Cut

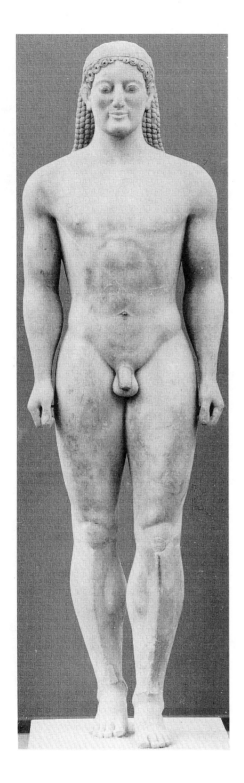

of the overall length of the body. For the Egyptian canon of proportions, the vertical axis passed through the navel. It is associated with the concept of the center, the point at which there is stillness and beginning. Our human origins take place in the womb where we are linked to our mothers by the umbilical cord, anchored at the navel.

This page:
Albrecht Durer. The Fall of
Man. *1504. Engraving. Courtesy of The Fogg Art Museum.*
Cambridge, Massachusetts. Gift
of William Gray from the Collection of Francis Calley Gray.

The figures exhibit Durer's interest in ideal proportions.

Given $\overline{TA}$

T ——————— A

1. Extend $\overleftrightarrow{TA}$ through A.

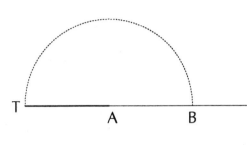

2. With the compass, place the metal tip on A and the pencil tip on T and cut a semicircle. Label the point of intersection B.

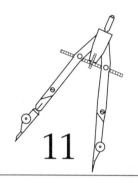

11

3. Use Construction 10 to find the Golden Cut of $\overline{AB}$ and label it C.

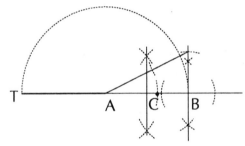

Divide a Line Segment Into Ø Proportions

4. Place the metal tip on C and the pencil tip on A and cut a semicircle. Label the point of intersection R.

5. Place the metal tip on B and the pencil tip on C and cut a semicircle. Label the point of intersection S.

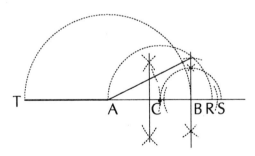

Now $\dfrac{TA}{AC} = \dfrac{AC}{CB} = \dfrac{CB}{BR} = \dfrac{BR}{RS} = \emptyset$

Note: A close approximation of Ø-proportion subdivisions can be obtained using graph paper marked off in consecutive Fibonacci numbers. It is a little easier to do. We refer the reader to Chapter 5 for further information.

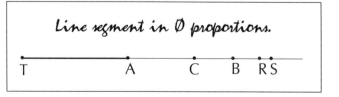

Line segment in Ø proportions.

T   A   C   B   R S

The Pentagon

The **pentagon** is introduced at this time rather than later on with the other polygons because it is the structure par excellence found throughout the animate world. The pentagon is a five-sided **simple closed curve** composed of line segments that intersect only at their endpoints. It is called regular if all sides and angles are congruent. It provides the framework for the **pentagram** which clearly demonstrates the Golden Ratio. It is rich with visible harmonic properties.

This page:
Natural forms exhibiting pentagonal structure.

Counterclockwise from top: Sand dollar, buttercup, apple, and starfish.

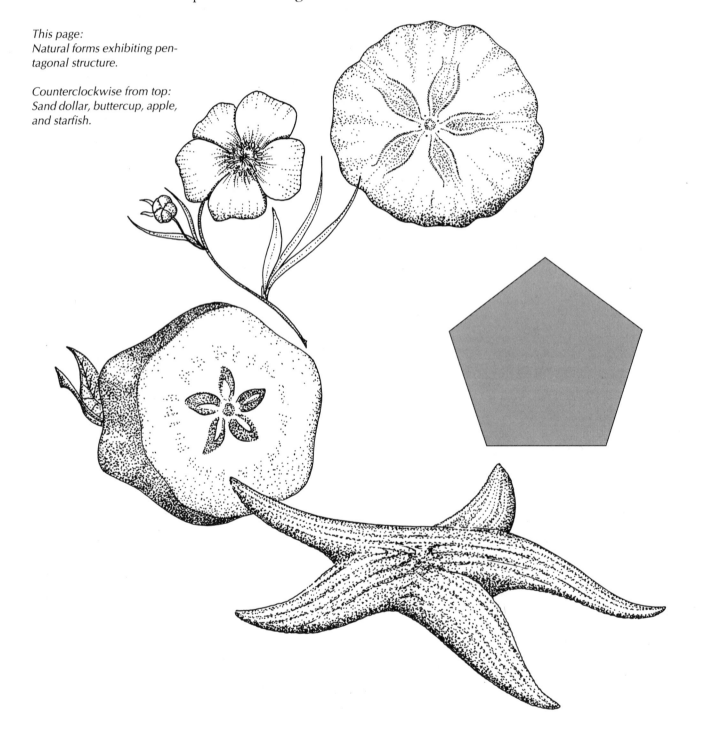

The power of the Golden Section to create harmony arises from its unique capacity to unite the different parts of a whole so that each preserves its own identity, and yet blends into the greater pattern of a single whole. The Golden Section's ratio is an irrational infinite number which can only be approximated. Yet such approximations are possible even within the limits of small whole numbers. This recognition filled the ancient Pythagoreans with awe: they sensed in it the secret power of a cosmic order. It gave rise to their belief in the mystical power of numbers. It also led to their endeavors to realize the harmonies of such proportions in the patterns of daily life, thereby elevating life to an art.

Gyorgy Doczi
The Power of Limits

Above:
Andrea Hart. Student Work.
Chapter 2   Project 2. Ink and cut paper on illustration board.

Below:
Pentagonal shape in a tropical plant.

The pentagram, also known as **pentalpha, pentacle** or **pentangle,** is the five-pointed star derived from connecting the **non-consecutive vertices** of a regular pentagon. Fig. 2.9 illustrates the appearance of the Golden Ratio in the different segments of the pentagram. Also, the segments in Golden Ratio are shared by the pentagram and its corresponding pentagon.

The pentagram is a fitting symbol of visually expressed harmony and order. It was the chosen symbol and seal of the Order of the Pythagoreans, a secret society of Ancient Greek mathematicians organized in the 6th century B.C. They were fascinated by natural harmony and order and believed it could be explained by numerical relationships.

There are several ways to construct a regular pentagon. The first of the three that follow is derived directly from the properties illustrated in Fig. 2.9. The other two are included since the given conditions for the problem may vary. It is one thing to generate a pentagon if the length of a side is known and quite another to fit a pentagon into a given space.

Right:
Fig. 2.9 Golden Ratio relationships in the pentagram and in the segments of the pentagram and its corresponding pentagon. In each of the figures:

$$\frac{AB + BC}{BC} = \frac{BC}{AB} = \varnothing \approx 1.61803$$

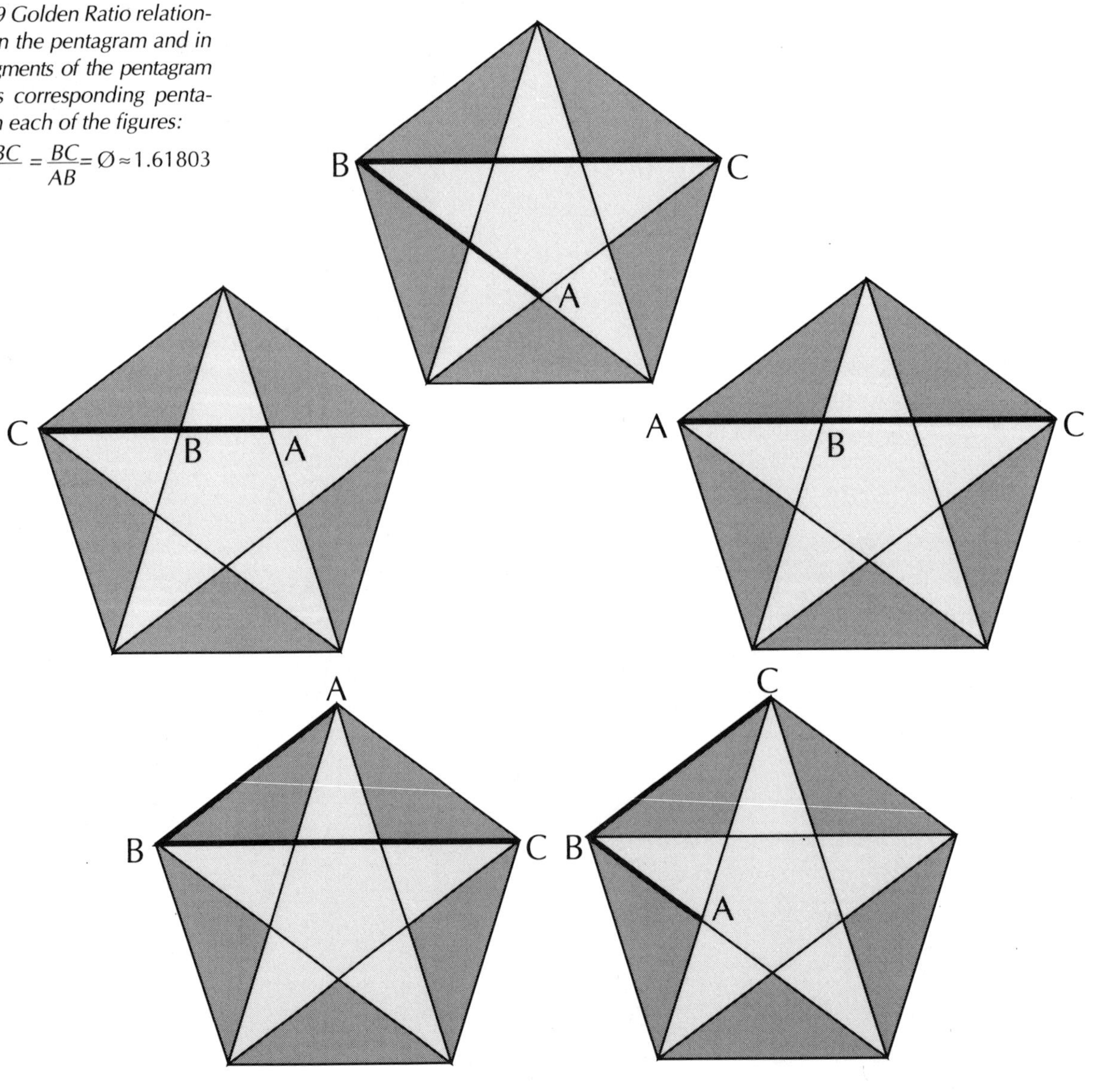

| Given C is the Golden Cut of $\overline{AB}$ |

1. Open the compass to measure CB. With the metal tip on C, cut an arc above $\overline{AC}$.

2. Without changing the setting, place the metal tip on A and cut an arc that intersects the one drawn in Step 1. Label the point of intersection D.

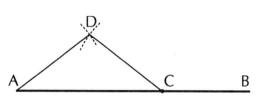

3. Draw $\overline{AD}$ and $\overline{DC}$. A, C and D will be three vertices of the regular pentagon. We need to find the other two.

4. Open the compass to measure AC. With the metal tip on A, cut an arc below C in roughly the position of the fourth vertex. Without changing the setting, and with the metal tip on C, cut an arc below A in roughly the position of the fifth vertex.

5. Open the compass to measure AD. First with the metal tip on A and then with the metal tip on C, cut arcs that intersect those drawn in the previous step. Label the points of intersection E and F.

6. Draw $\overline{AE}$, $\overline{EF}$ and $\overline{CF}$.

Now AEFCD is a regular pentagon.

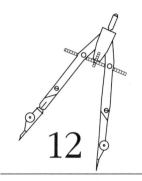

12

Construct a
Regular
Pentagon
Given the
Golden Cut of
a Line Segment

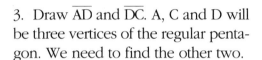

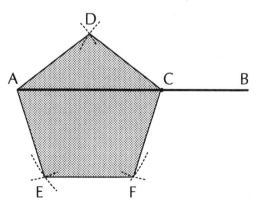

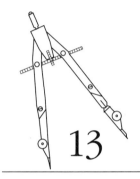

13

Inscribe a
Regular
Pentagon in a
Circle

*Make sure your
compass lead is
sharp to insure
accuracy. This
construction can
be very frustrat-
ing!*

Given Circle O

*When you draw the circle, the metal
tip of the compass will determine the
center of the circle.*

1. Draw a **diameter** and label its endpoints A and B, respectively.

2. Construct another diameter perpendicular to $\overline{AB}$. Label the endpoints C and D, respectively.

3. Bisect $\overline{OB}$ and label its midpoint E.

4. Put the metal tip on E and the
pencil tip on C and cut an arc that
intersects $\overline{AB}$. Label the point of intersection F.

5. Place the metal tip on C and the
pencil tip on F and cut an arc that
intersects the circle. Label the point of
intersection G.

6. Draw $\overline{CG}$.

7. Without changing the compass setting, put the metal tip on G and cut
another arc that intersects the circle.
Label the point of intersection H.

8. Draw $\overline{GH}$.

9. Without changing the setting, continue around the circle placing the
metal tip on H to cut intersection I,
and on I to cut intersection L.

10. Draw $\overline{HI}$, $\overline{IL}$, and $\overline{CL}$.

Now CGHIL is a regular pentagon.

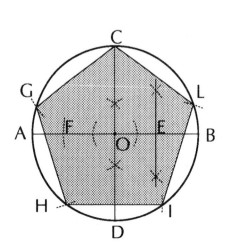

This page:
Pentagon and pentagram variations.

Top:
Nadine Beckwith. Student Work. Chapter 2 Project 2. Photocopier elements and black marker.

Bottom:
George Reyes. Student Work. Chapter 2 Project 2. Pen and ink on paper.

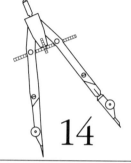

14

Construct a
Regular
Pentagon
Given a Side

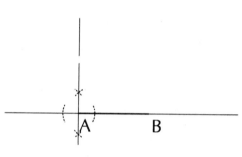

Given $\overline{AB}$

1. Extend $\overleftrightarrow{AB}$ through A and B.

2. Construct a line perpendicular to $\overleftrightarrow{AB}$ at A.

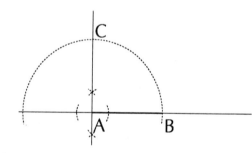

3. Put the metal tip on A and the pencil tip on B and cut an arc from B well through the line perpendicular to $\overleftrightarrow{AB}$. Label C the point of intersection of the arc and the line perpendicular to $\overleftrightarrow{AB}$.

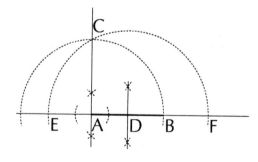

4. Bisect $\overline{AB}$ and label the midpoint D. Put the metal tip on D and the pencil tip on C and cut an arc that intersects $\overleftrightarrow{AB}$ on both sides of D. Label the points of intersection E and F.

5. Put the metal tip on A and the pencil tip on F and cut an arc on the same side of $\overleftrightarrow{AB}$ as arc ECF. Without changing the setting, place the metal tip on B and cut an arc that intersects the one just drawn. Label G the point of intersection. Label H the point of intersection of arc EG and the arc drawn in Step 3.

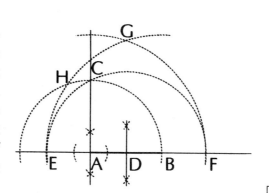

6. Put the metal tip on B, the pencil tip on A and cut an arc that intersects arc GF. Label I the point of intersection. Draw $\overline{AH}$, $\overline{HG}$, $\overline{GI}$, and $\overline{IB}$.

Now AHGIB is a regular pentagon.

Make sure your compass lead is sharp to insure accuracy. Minor inaccuracies at the beginning get magnified as work progresses.

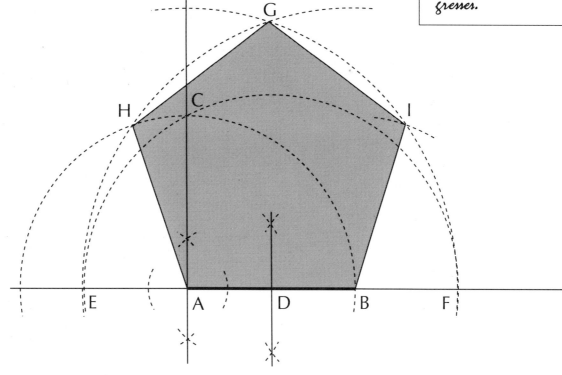

This page:
Linda Maddox. Student Work.
Variations on the theme of a
pentagon. Technical pen and
ink on paper.

The **decagon,** or ten-sided polygon, can be easily constructed from a pentagon inscribed in a circle. If the perpendicular bisector of one side ($\overline{QP}$) is drawn, the point at which it intersects the circle (R) bisects the arc cut off by the side of the pentagon. The remaining arcs may be bisected by measuring RP with the compass and using this measurement to mark off subsequent arcs around the circle. Connect these points to be the vertices of the decagon. The decagon is useful since it inherits the harmonic properties of the pentagon, and it offers diverse and interesting design possibilities.

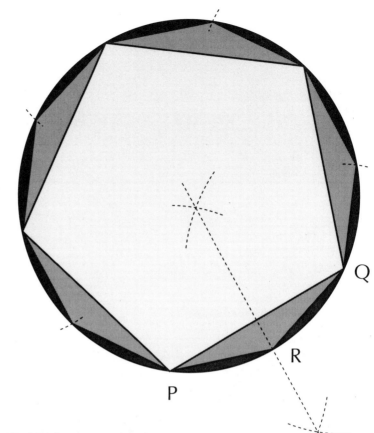

Above:
Fig. 2.10 Construction of a regular decagon is achieved by bisecting the sides of an inscribed regular pentagon.

Left:
An example of a natural decagon is evident in the head of a passion flower.

Top :
Fig. 2.11 Each of these may be derived from the intersecting inscribed pentagons shown in the first design. Designs c, d, and e are variations of design b, gotten by eliminating line segments and shading appropriately.

Bottom left:
Fig. 2.12 Diagram related to the student work.

Bottom right:
Lana Yonkers. Student Work. Chapter 2 Variation on Project 2.

An effective variation on the decagon is shown in Fig. 2.12. The line diagram illustrates the principle that was used in the accompanying student work. The points are first marked on the circle. Then, line segments are drawn connecting every third point.

Since ten divided by three yields three with a remainder of one, every three consecutive segments begin and end one unit apart. By continuing the process, each point on the circle is used up and eventually the path returns to the starting point.

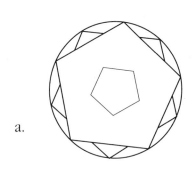

a.

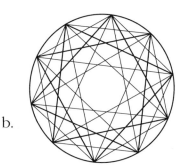

b.

c.

d.

e.

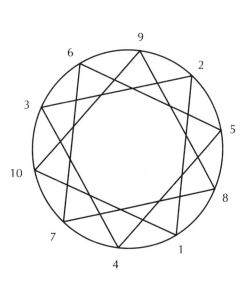

The Golden Rectangle

We have examined the existence of the Golden Ratio, Ø, in the pentagon and the pentagram and now turn our attention to the rectangle that has become known as the **Golden Rectangle.** If the lengths of the sides of a rectangle are in the Golden Ratio, then the rectangle is a Golden Rectangle.

The mathematician has explored this rectangle as an abstract intellectual exercise. The artist has used it as the structure for harmonious compositions. And the Golden Rectangle is an implied "frame" for many natural forms.

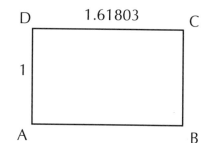

Top:

Fig. 2.13 $\dfrac{DC}{DA} = Ø \approx \dfrac{1.61803}{1}$

Below:
Natural creatures set within implied frames.
Center left:
Moth
Center right:
Sea Turtle

Bottom left:
Common Black Hawk

Bottom right:
Sergeant Major fish

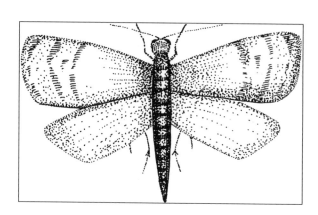

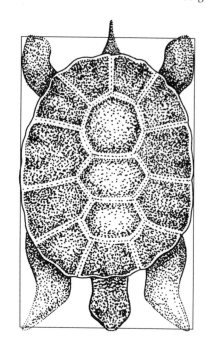

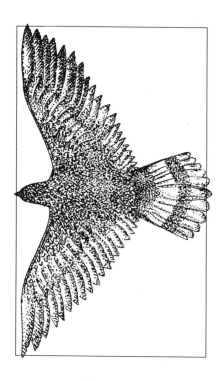

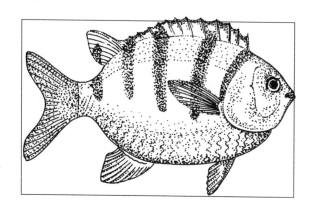

This page:
Fig. 2.14 Golden Rectangle.

Top to bottom:
ABCD and its unit square.

Successively smaller Golden
Rectangles.

A larger Golden Rectangle.
$$\frac{DE}{EF} = \frac{2.61803}{1.61803} \approx 1.61803 \approx \emptyset$$

ABCD (Fig. 2.14) is a Golden Rectangle. We have assigned the numbers 1 and 1.61803 to the sides of this rectangle independent of real world measurement standards such as inches, meters, etc. The width becomes the unit measure regardless of the size of the rectangle; therefore any Golden Rectangle may be said to contain a **unit square.**

If we cut off this square from the parent rectangle, what remains is rectangle BCEF with dimensions of 1 and .61803. Since

$$.61803 \approx \frac{1}{\emptyset}; \quad \frac{1}{.61803} \approx \emptyset$$

BCEF is also a Golden Rectangle and is similar to ABCD. If we then cut a square from BCEF, we are again left with another smaller Golden Rectangle since

$$\frac{GH}{HC} = \frac{.61803}{.38197} \approx 1.61801 \approx \emptyset.$$

We could continue with this process indefinitely getting smaller and smaller Golden Rectangles.

To increase the size of our original rectangle, then, we need only construct a square on the side measuring 1.61803. We could continue with this process indefinitely getting larger and larger Golden Rectangles.

The Golden Rectangle is the only rectangle from which a square can be cut and the remaining rectangle will always be similar to the original rectangle.

If you have run out of Golden Rectangle mixes, the next construction gives a recipe for whipping up one from scratch!

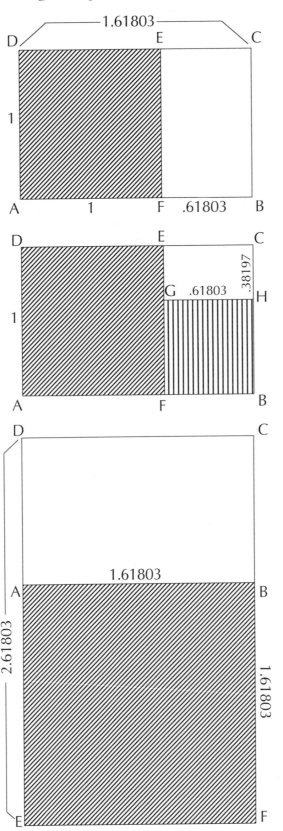

Given $\overline{AB}$

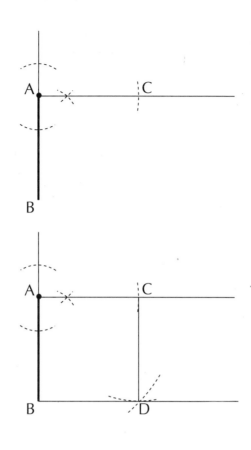

1. Extend the line through A and construct a perpendicular to $\overleftrightarrow{AB}$ through A.

2. Open the compass so that it measures AB. Without changing the setting, place the metal tip on A and cut an arc on the perpendicular constructed in Step 1. Label the point of intersection C.

3. Place the tips of the compass on B and C respectively. Without changing the setting, put the metal tip on A and cut an arc in the interior of ∠BAC.

4. Put the metal tip on C and the pencil tip on A and cut an arc that intersects the one drawn in Step 3. Label the point of intersection D.

5. Draw $\overline{BD}$ and $\overline{DC}$ to form a square. Extend $\overrightarrow{BD}$.

6. Bisect $\overline{BD}$ and label its midpoint E.

7. Put the metal tip on E and the pencil tip on C and cut an arc that intersects $\overleftrightarrow{BD}$. Label the point of intersection F.

8. Place the tips of the compass on B and F respectively. Without changing this setting, put the metal tip on A and cut an arc on $\overrightarrow{AC}$. Label the point of intersection G and draw $\overline{GF}$.

Now ABFG is a Golden Rectangle generated from the square ABDC, and

$\dfrac{BD}{DF}$ is the Golden Ratio.

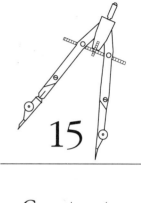

15

Construct a Golden Rectangle Given the Width

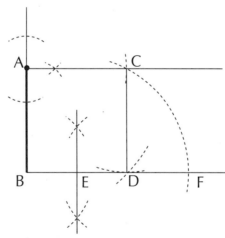

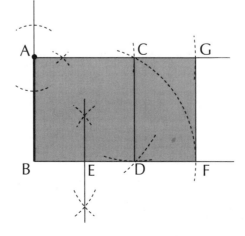

The Golden Rectangle provides a structure on which to build a proportional Golden Rule. Using the Rule, you can easily find the Golden Cut of any line segment whose length is less than that of the Golden Rectangle.

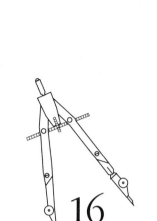

16

Construct a Golden Rule

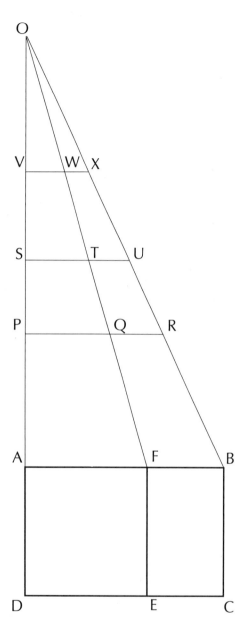

Given Golden Rectangle ABCD with its unit square ADEF

1. Extend $\overrightarrow{DA}$ and choose a point O on it that determines the length of the Rule.

2. Draw $\overline{OF}$ and $\overline{OB}$.

Now any line segment drawn parallel to $\overline{AB}$ in Δ OAB will be divided at the Golden Cut by $\overline{OF}$.

eg. $\dfrac{PQ}{QR} = \dfrac{ST}{TU} = \dfrac{VW}{WX} = \emptyset$.

Here is a useful construction for increasing or decreasing the sides of a rectangle without destroying its proportions. It works, not only for the Golden Rectangle, but for any rectangle, and will be particularly useful for those explored later on in Chapter 4. When completed you will notice that this construction would be simplified tremendously by using a T-square and triangle!

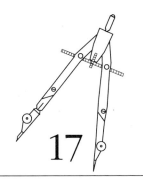

17

Construct a Rectangle Similar to a Given Rectangle

Given rectangle ABCD

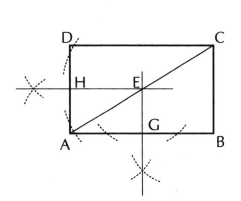

A. To construct a smaller rectangle:

Draw diagonal $\overline{AC}$. Choose an arbitrary point on $\overline{AC}$ and label it E. Through E, construct perpendiculars to $\overleftrightarrow{AB}$ and $\overleftrightarrow{AD}$. Label points of intersection G and H.

B. To construct a larger rectangle:

Extend $\overrightarrow{AC}$. Extend the sides of the original rectangle through B and D. Choose an arbitrary point on $\overrightarrow{AC}$ in the exterior of ABCD and label it F. Construct perpendiculars through F to $\overleftrightarrow{AB}$ and $\overleftrightarrow{AD}$. Label points of intersection I and J.

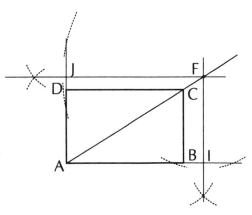

Now AGEH and AIFJ are similar to ABCD.

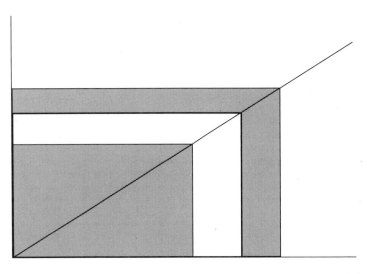

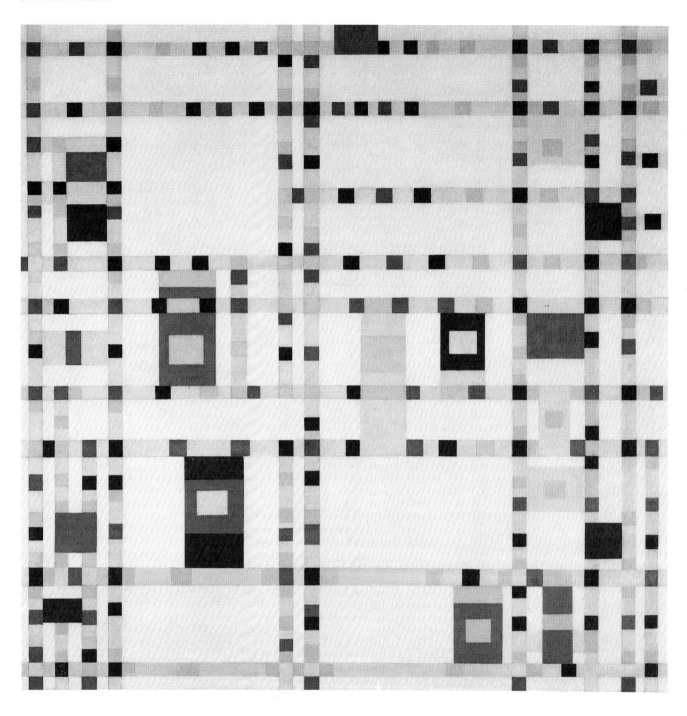

Above:
Piet Mondrian. Broadway Boogie-Woogie. *1942-43. Oil on canvas. 127cm. x 127cm.; O464, S423. Collection, the Museum of Modern Art. Given anonymously.*
This painting abounds in Golden Ratio subdivisions in both the vertical and horizontal lines.

It is essential for students of the arts to understand and be able to use harmonious relationships in their compositions. Art is, in part, an intellectual discipline. The intellect determines the underlying structure for a specific work and the emotions provide the content.

Two of the great periods of art, Classical Greece of 5th century B.C. and Renaissance Italy of the 15th and 16th centuries A.D., the latter being indebted to the former, sought an understanding of the world of mathematics so that their arts would have a sound structural base. Greek sculptors and architects, as well as Renaissance artists, relied heavily on the Golden Rectangle for the armature of their works. Twentieth century artists, such as the architect Le Corbusier, and the Dutch painter, Mondrian, have also been inspired by principles found in geometry. In the book, *The Painter's Secret Geometry,* author Charles Bouleau examines a great many paintings in Western art in order to understand the mathematical structure beneath the outward appearances. We refer you to this book for explanations of the works of artists such as Duerer, Da Vinci, Vermeer, Seurat, and Mondrian.

It is an impressive discovery when the human mind first catches glimpse of the eternal supersensuous laws ruling the seemingly casual appearances of the world of sense. This moment came to the Greeks early in their career in the course of Pythagorean and other geometric investigations.

Rhys Carpenter
The Esthetic Basis of Greek Art

This page:
Linda Maddox. Student work.
Design developed within a
Golden Rectangle.

The Parthenon

Let us examine, now, one of the most famous pieces of architecture produced in the Western world. We shall see how the Golden Rectangle is an integral part of the form. But, first, let us give you some background on that structure.

On the highest part of the acropolis in Athens, Greece, standing among the other temples, is the Parthenon. It was built during the years 447-432 B.C., and is still there today despite the ravages of time and human enemies. The temple was built under the aegis of Pericles during the Golden Age (450-400 B.C.). Ictinus and Callicrates, the architects, and Phidias, the sculptor, were responsible for the aesthetics of the work. Its function was to serve both as a treasury and a dwelling house for the goddess Athena.

Below:
Le Parthenon, facade, cote Est,
Athenes, Grece Acropole.

The Greeks are an outdoor people. They relate to sea and sky and mountains under a very brilliant light, and the Parthenon was meant to be seen from many vantage points around the city. Every four years there was a festival held in honor of Athena to whom the temple was dedicated. She became the protectress of the city when she won a contest against the god Poseidon. By prodding the ground with her lance, she brought forth an olive tree as her gift to humanity. The sculptural elements on the building give the details of this contest as well as other aspects of her story.

In order to reach the Parthenon, the worshippers had to move along the Panathenaic way, the main avenue of the city, go up a difficult steep climb, and enter a gateway before coming upon the temple. Then, the celebration procession would begin on the west side of the building, run along the north and south sides in two separate groups, and converge at the center of the east front. Here was concluded the "ceremony of the robe" in which a new cloak of saffron and purple was placed on the statue of Athena. According to Duane Preble in his book, *Man Creates Man*, the axis of the building was carefully calculated so that, on the birthday of Athena, the rising sun would come through the east doorway to fully illuminate the gold covered statue created by Phidias.

Below:
Reconstruction drawing by Sylvia T. Burnside showing a view of the Parthenon on the acropolis.

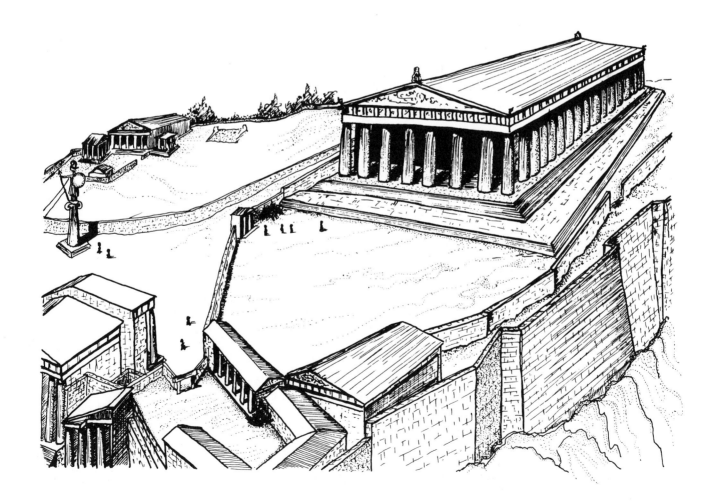

This building is the culmination of a long evolution of the particular Doric style of architecture, the simplest of the three classical modes. The other two are called the Ionic and Corinthian, being named after the type of ornamentation at the top of the columns. The Parthenon is a special refinement on the essentially simple post-and-lintel method of construction. It is a rectangular marble structure that is 228' long by 104' wide with columns 34' in height. It was a totally marble creation except for the timbered roof under the marble tiles and the wooden doors and frames, long gone from this building. While the exterior rectangle was composed of a series of columns, the interior was a windowless room which housed both the treasury and the cult statue of Athena. At the short ends of the temple were eight columns and on the long sides there were seventeen. There were porticos on both front and back with a walkway between the colonnade and the **cella**, though essentially the building was meant to be used as exterior space.

In this piece of architecture, there is an orderly organization of the vertical members to the horizontal ones; the length and breadth to the height. The ratio of solid masses of columns to open voids was calculated by a unit of measure known as the module. This module was developed by an internal unit such as the radius or diameter of a column, not by an outside unit such as feet or inches.

Now let us take a look at a rather surprising element in the design of this temple. Despite the concern for geometry, the structure hardly possessed a straight line. The desire seemed to be to create a sense of the organic, filled with life, rather than to produce a mechanical, only mathematically "correct" system.

Though the stepped courses at the base of the building appear to be equal in height, they are not. The lowest is the narrowest and the stylobate, the topmost, is the widest. The entire platform is not evenly horizontal but rises convexly in the center. The stylobate arches slightly from corner to corner having the center point on the short

Below:
Fig. 2.15 Detail of Facade of the Parthenon.

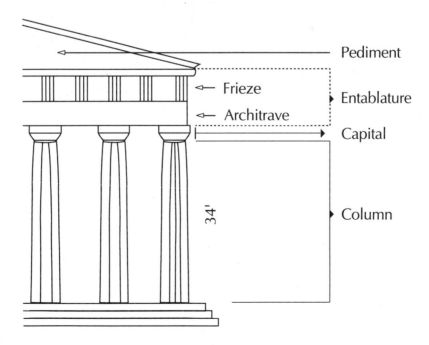

ends about 2.75 inches higher than the corners, and the long sides rise about 4 inches in the center. According to some references, the reason for these deviations seems to be tied to optical perceptions. If the measurements were even, the platform would appear to be hollow when seen from the side. Above the column, the architrave echoes this curve. The entire base, therefore, optically resists the downward pull of the monument.

The columns have a slight swelling as they rise, .69 of an inch, to give the eye the sense of straining, as if these were human muscles meant to carry the heavy load of the entablature. This swelling is called entasis. While the columns themselves appear equally spaced and all perpendicular, they are not. The ones at the four corners are nearer to their neighbors than the others. This is because they are seen against open space rather than against the solid mass of the inner cella wall. They also serve as members of both the lateral and longitudinal colonnades. The columns are also 2" thicker here. They all incline to varying degrees toward the center, some off plumb by 3", so that the whole appears to converge at some point in the heavens. All of these suggest a conscious and considered reaching upward.

Greek architecture is an art of related lines and surfaces on the plane, not of solids in space. Instead of true depth, what we have is successive and usually parallel planes in space, a presentation of two dimensions rather than felt volumes existing in space. As a result, only one angle, the right angle of 90 degrees, is permitted. Since this angle can only appear in parallel or perpendicular planes, we can only define or bound solid space, but we cannot enclose it and have

that enclosed space be perceived as a felt "something." We needed to wait until the twentieth century before the architect could consciously be concerned with "emptiness."

Unlike the civilizations that went before theirs, or even with those that were contemporaneous, the Greeks believed that Nature was orderly, and that her laws could be comprehended rationally. With this came a belief that there is an ideal form for every class of objects and that such a true

Below:
Fig. 2.16 Floor plan of the Parthenon.

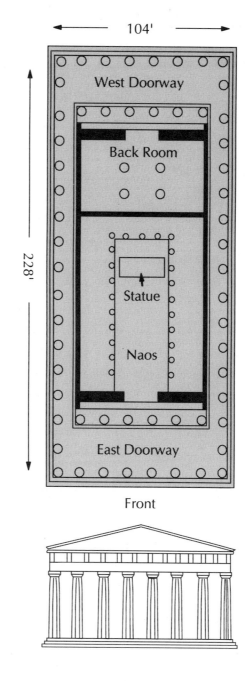

67

form is characterized by its geometric simplicity and by the **commensurability** of its component members. This idea could then be expressed through Art. Thus, the Parthenon is a rational structure and geometrically comprehensible. It is believed that it was designed "on paper" before it ever was built. We can study its proportions in terms of the original design and not be concerned with the departures resulting from special circumstances that might be due to time or geography or natural forces.

Using a geometric basis for design would allow a building to be conceived and reviewed before beginning construction. The proportions would be accurate at any scale and what would be perceived is the essential integrity of the structure without overattention to detail. With the use of compass and straightedge, the full scale structure could then be laid out on the ground with the help of some sort of measuring tape. Thus, the actual dimensions of every member of the building could be accurately ascertained and then recorded in the particulars of Greek feet and their subdivisions. This would be an indispensable tool for writing the necessary specifications for quarrying, finishing and setting in place the actual blocks of marble.

The Greeks, as we shall see in subsequent chapters, recognized a geometric commensurability. What would not be commensurable in linear units would be in units of area. We are indebted to Jay Hambidge, author of *The Parthenon and Other Greek Temples,* for his work in analyzing the Parthenon. The theory of **root rectangles**, which we shall examine in Chapter 4, is the essence of the theory of proportion that he used in his analysis, especially the principle of proportion inherent in the root-five rectangle (a relative of the Golden Rectangle). There seems to be a constant recurrence in Greek works of art of a very limited number of proportions which are derived by simple constructions beginning with the square. The proportions that Hambidge found could have been laid out with the help of pegs and cords, as we shall see when we look at Egyptian architecture in the following chapter.

In Fig. 2.17 we offer an analysis of the facade of the temple and its relationship to the Golden Rectangle. In Chapter 5 we shall examine the Parthenon again in terms of the other **Dynamic Rectangles.**

On the preceding pages we have begun to examine the role of $\emptyset$ as it relates to the disciplines of mathematics and art through the Golden Rectangle. In his book, *The Power of Limits,* Gyorgy Doczi has looked in detail at many natural forms and has found that certain proportions relating to $\emptyset$ appear again and again. He analyzes such diverse creatures as butterflies, fish, beetles, skates, and humans. We refer the reader to this book for an in-depth study. In the following chapters we shall continue to examine $\emptyset$ as a common thread that weaves through the patterns of our existence.

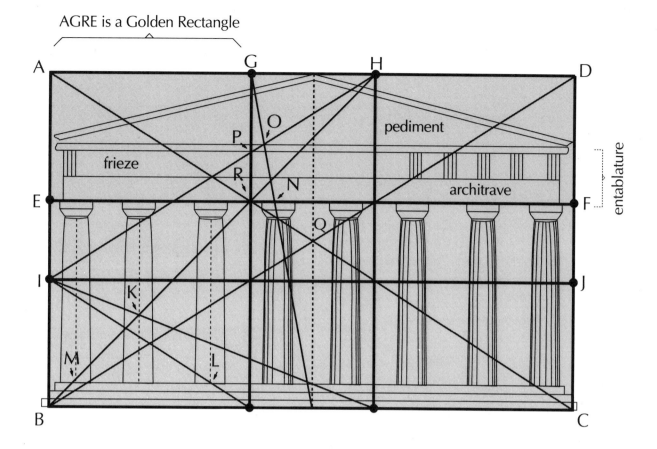

AGRE is a Golden Rectangle

Fig. 2.17 Analysis of the Facade of the Parthenon

ABCD is a Golden Rectangle.

E and F are the Golden Cuts of $\overline{AB}$ and $\overline{DC}$ respectively, and $\overline{EF}$ underscores the architrave.

The pediment and the entablature can be subdivided by Golden Cuts G and H into two Golden Rectangles and a square.

By further dividing the vertical sides with Golden Cuts I and J and drawing in the diagonals of the various rectangles formed (called an harmonic subdivision of the rectangle), we see that the columns are centered on points of intersection of the diagonals alone or the diagonals and sides of the smaller rectangles (see points K, L, M, and N).

The placement of the frieze and the pediment are also determined by such points of intersection (see points O and P).

Since there is symmetry about the vertical line through Q, corresponding points would be determined on the right hand side of the facade by drawing in the corresponding diagonals on that side.

Problems

Right:
Michelle Gogas. Student work.
Chapter 2   Project 5. Water-
color on illustration board.

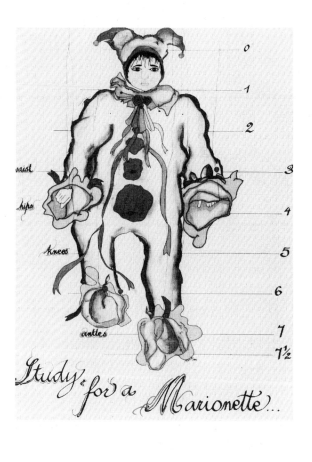

1 How do you measure up to classical standards of beauty? In the sculptures pictured in the text, body measurements appear in the Golden Ratio. The classical sculptors frequently used the navel as the Golden Cut of the body. Is your navel the Golden Cut of your body?

A. Floor to navel = _____

B. Overall height = _____

$\dfrac{B}{A}$ = _____

2 Research body proportions in the figurative sculpture of a nonwestern culture. In writing and with visual support, compare and contrast your findings with the Classical Greek model.

3 Construct a Golden Rule (Construction 16) from a rectangle whose length is at least as long as your middle finger. Use the Rule to analyze the proportions of your hand.

4 Find the Golden Cut of the line segments to the right. Use either segment to construct a pentagon.

5 Divide the line segment to the right into at least 4 segments in Ø proportions. Disregard any leftover portion of the line segment.

6 a. Inscribe a figure, similar to the one on the right, in a circle whose radius is a minimum of two inches.

  b. Construct a similar figure without using the circle.

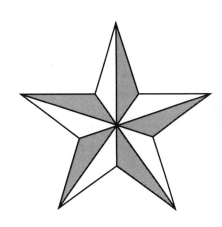

7 Using your newly acquired information about Golden Rectangles, do an analysis of common rectangular objects around the house. Examine such things as playing cards, stationery, magazines, books etc. Draw up a chart giving dimensions, type of article, and ratio of length to width of the rectangles. Which ones are close to Golden Rectangles, which are not?

8 Construct four Golden Rectangles of different sizes.

9 Reconstruct one of the figures on page 56.

10 a. Find a front view of either a male or female face in a magazine. Analyze the facial structure using Fig. 2.8 as a guide.

  b. Repeat, using a photo of an entire human form.

Projects

1 Using an illustration board or heavy paper, divide a line segment of any length into Ø proportions *(see Construction 10)*. Use this line segment as a "ruler" to first construct a rectangle and then to subdivide its interior in some way into Ø proportions. Color in a manner of your choosing.

2 Using the pentagon and pentagram variations, *(refer to Constructions 12 - 14)*, create a pattern in black and white that would be suitable for fabric or wrapping paper design. Repetitions of your pattern may be achieved by: stencil, silk screen, linoleum cut, woodblock, rubber stamp, potato print, photocopy, computer, etc. Your finished work should measure at least 8.5" x 11", and may be done with the materials of your choice.

3 Using a maximum of four colors, and Golden Rectangles and squares, create a masterpiece (!) in either two or three dimensions.

4 Using the above photographs of life forms for inspiration, create your own work in two or three dimensions. Consider the possibilities of a mobile, a quilt, a banner, a poster.

5 Make a marionette using the classic concept of body proportions involving a total length of 7.5 heads as described at the beginning of this chapter. It may be as simple as a jointed figure made of paper (such as a shadow puppet) or as complex as a stringed wooden figure. Your solution will depend on time constraints as well as available materials. Consider doing a self-portrait using old photographs and a collage approach.

6 a. *Found Face:* Using found photos of facial features (phew!), create two collages; one in which there is harmony, and one that breaks the harmony to tend toward the grotesque. This could also be a basis for a mask.

b. *Found Figure:* Use the same concept to create two different human body forms.

Below:
Kay Holloran. Student Work. Chapter 2   Project 7. Ink on watercolor paper.

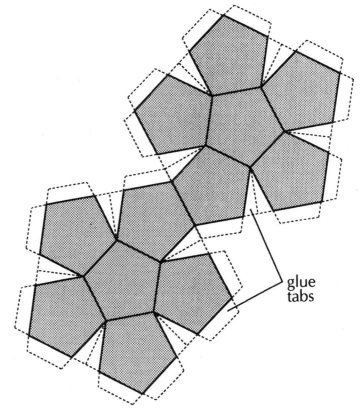

Right:
Two-dimensional pattern tem-
plate for the construction of a
three-dimensional form called
the dodecahedron.

glue
tabs

7 Sculptural Form

Twelve pentagons join together to form a three-dimensional figure called the **dodecahedron**. The pattern below shows how to lay out the twelve pentagons in the plane. By pulling the pentagons together so that edge meets edge, the dodecahedron is formed. Create your own variation on this **Platonic Solid** by:

a. Making a template out of heavy paper following the directions in Construction 11 or 12. Use this template to make all 12 units.

b. Divide each pentagonal face into interesting subdivisions and fill in with black and white or color.

c. Join all 12 units, or multiples of 12, to create a sculpture.

d. Use materials of your choice to make the finished form. Consider the possibilities of using:

1. felt material
2. patchwork units
3. knitted units
4. crocheted units

(The above would need to be stuffed to fill out the units.)

5. Fome-Cor board
6. Plexiglas
7. sheet copper, brass or aluminum
8. colored acetates
9. stained glass pieces
10. clay units
11. any other materials that you might fancy.

Further Reading

Bouleau, Charles. *The Painter's Secret Geometry*. New York: Harcourt Brace and World, 1963.

Cook, Theodore Andrea. *The Curves of Life*. New York: Dover Publications, Inc., 1979.

Doczi, Gyorgy. *The Power of Limits*. Boulder, Colorado: Shambala Publications, Inc., 1977

Ghyka, Matila. *The Geometry of Art and Life*. New York: Dover Publications, Inc., 1977.

Grillo, Paul Jacques. *Form, Function, and Design*. New York: Dover Publications, Inc., 1960.

Hambidge, Jay. *The Parthenon and Other Greek Temples*. New Haven: Yale University Press, 1924.

Huntley, H.E. *The Divine Proportion*. New York: Dover Publications, Inc., 1970.

Robertson, Martin. *A Shorter History of Greek Art*. Cambridge: Cambridge University Press, 1987.

Seuphor, Michel. *Piet Mondrian, Life and Work*. London: Thames and Hudson, 1957.

Woodford, Susan. *The Parthenon*. Cambridge: Cambridge University Press, 1986.

3 *Special Triangles*

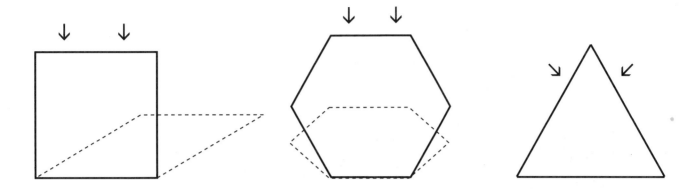

Mathematically both the terms *plane* and *space* are undefined. For all practical purposes, humans experience and understand three-dimensional spatial relationships intuitively. However, artists and mathematicians need to approach the two-dimensional surface, that "slice out of space", more thoroughly and consciously. In order to investigate the plane, it is helpful to give boundaries to parts of it.

The simplest figure composed of straight line segments that encloses a portion of the plane is the triangle (see Fig. 3.1). Neither one nor two line segments is able to do this. Structurally, it is not only the simplest but also the most stable form as it is unable to be collapsed. It is a wonderfully economical figure and certainly one with a long history of use both as a foundation for construction and a tool for examination.

In the Ancient World, especially for the Egyptian civilization, the need for bounding a portion of the plane was a problem of extreme importance. Each year, the Nile River overflowed its banks erasing the boundaries which marked the rich farm lands adjacent to the water's edge. Some method was needed to re-establish those boundaries. Thus, for the Egyptians, it appears that the use of geometry existed for purely practical purposes long before the Greeks gave it the name meaning "earth measure." The Egyptians needed to have a working knowledge of **points** (boundary markers), **lines** (distances between markers) and **areas** (the fields enclosed by the boundaries). They had discovered how to find the areas of square, rectangular and triangular shapes. Not all fields, however, lent themselves to these simple regular shapes. If, though, the sides of the fields were straight, any one could be reduced to a series of triangular shapes whose areas could be found and then added together to give the total area. We call this process **triangulation**, a technique still used by surveyors today. Through trial and error, and the use of **inductive reasoning,** the Ancients acquired knowledge of some geometrical principles and began to develop simple working tools.

Above:
Fig. 3.1 The triangle is the only polygonal figure unable to be collapsed.

This knowledge and these tools were also needed for the construction of those architectural marvels, the pyramids. If we consider the Parthenon to be a building erected on a scaffolding of reason, then the pyramids can be seen as sarcophagi enveloped in mystery. This is especially true of the three at Giza on the west bank of the Nile River, the traditional place of the dead. These are known as the *Great Pyramids* of Menkaure, Khafre, and Khufu, and were built during the Fourth Egyptian Dynasty B.C. No one monument, or group of them, has teased the minds of so many.

These sophisticated cairns of stone brought the building of pyramids to the level of excellence. As royal burial tombs they were scaled to the dimensions of the gods as opposed to the size of man. While the Parthenon was accessible to all the citizens of Athens, the pyramids were accessible only to the god-king, the pharaoh, and his priests.

For what purpose were they built? There is widespread belief that the structure was an elaborate burial tomb of royalty in order to guarantee that in the afterlife the pharaoh would have all that he required and enjoyed on Earth. Some scholars, however, believe that pyramid construction was a complex social program that would bring the people together for a common purpose, one that was in keeping with political and economic factors rather than religious ones. Since humans live at the practical and symbolic level simultaneously, one reason does not negate the other.

Before the success of the three Great Pyramids, there was much trial and error in the form and the size of them. Different approaches were experimented with until perfection was achieved in the great one of Khufu (Cheops, to the Greeks). It has come to be the symbol of excellence on many levels. Built by the architect Imhotep (Asklepios, to the Greeks) and completed around 2,500 B.C., it was a costly use of resources for much of the 23 years that Khufu reigned. Its base is square, each side measuring 760'; height 482'; slope 51°52'. It orients to the four cardinal points of the compass (N, E, S, W), differing no more than 3'6" from the exact north-south orientation. It covers approximately 13.5 acres, which is equal to about 19.25 football fields. Its total weight is around 6 million tons. It, therefore, had to be erected on a very sturdy substrata of rocks.

The builders had to be able to measure in the air as well as on the ground. Once the square base was set (no small task for a structure so large!), the architects needed to be certain that

Below:
The Pyramids of Giza
Marburg/Art Resource, N.Y.

the building blocks were at the proper angles to the ground. A pyramid was composed of three thicknesses of blocks set in successively smaller square layers, each course being centered on the previous one. A single pyramid could have as many as 123 courses with many blocks weighing up to 2.5 tons each, and some weighing as much as 15 tons.

In laying out the base of the pyramid, the architects needed to be certain that the corners of each block were at right angles, since an inaccurate angle at a corner would throw the entire structure off. They probably used a method that closely approximates the technique for bisecting a line segment given in Construction 7. The difference was that the line segment used was a rope stretched taut between two markers, and the compass was a cord anchored at one end. Remember, the lever was the only engineering tool known to the builders of Ancient Egypt.

They discovered that if they cut a cord longer than half the length of the stretched rope and then swung arcs from both markers, the points where these arcs intersected would lie on a line at right angles to the original line, in this case, the stretched rope. In order to guarantee that the courses of massive stone blocks were perpendicular to the ground, the builders found that if they hung a heavy weight by a piece of cord and then let it swing until coming to rest, the right angle would be insured. This device is called a plumb line and is still in use today.

The blocks themselves had to be fitted together in such a way that the level of the course was preserved and the corners contained the proper angles. In each horizontal layer the

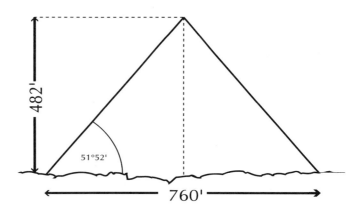

Above:
Fig. 3.2 The Pyramid of Khufu. This structure is situated on a square base covering approximately 13.5 acres and weighs about 6 million tons.

Below:
Fig. 3.3 Early Egyptian tools for construction.

Clockwise from top:
Level
Plumb line
"Compass"
Square

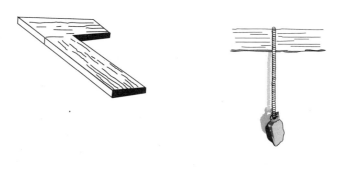

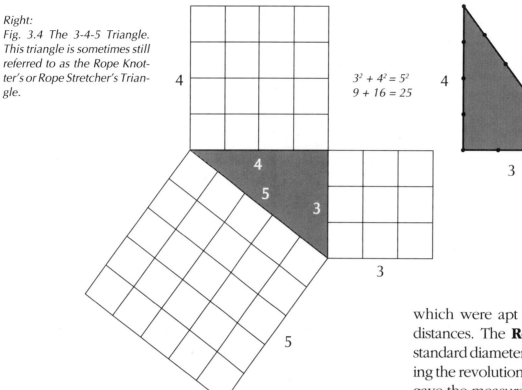

$$3^2 + 4^2 = 5^2$$
$$9 + 16 = 25$$

blocks were set in on a gentle grade so that those at the edges were higher than those in the middle. This made the whole layer slightly concave towards the center providing an inward thrust to counteract the lateral forces. The need arose for a consistent and accurate carpenter's square. Such a tool necessitates a right angle which is most easily obtained from a right triangle. It is believed that the discovery of the right-angled triangle was made by the rope knotters of early Egypt. Their job was to tie equally spaced knots in long ropes which would then be used for measuring. The distance between two consecutive knots was the unit measure, which had to be standardized for large-scale building.

Some linear measurements may have been taken by using a rolling drum rather than ropes of palm, which were apt to stretch over long distances. The **Royal Cubit** was the standard diameter of the drum. Counting the revolutions that the drum took gave the measurements.

Each pyramid was built in stages. First, a tall stepped structure had to be erected. A flag was placed at the top so that sightings could be taken throughout the building process. After that, packing blocks were laid on to insure the shape of the pyramid. Finally, an outer casing was added, starting at the bottom. It was completed with a smooth finish. Although the technology was simple, the results were impressive. The pyramids were so skillfully built that the joints of the massive blocks are scarcely visible. The damage done to them has been through the hands of man and not the normal erosion of time.

In stretching the ropes around three pegs, again the minimum number needed to enclose an area by straight sides, they discovered that a rope with exactly 13 knots (12 units) formed a triangle that had one right angle. Its sides could be measured in whole numbers, important for a culture (not so unlike ours!) that could not deal comfortably with fractions. The legs of this triangle measured 3 and 4 units, respectively, and the **hypotenuse** measured 5 units. Thus, it is called the 3-4-5 Triangle. It had the added property that the sum of the squares on the legs equalled the **square** on the hypotenuse, although it was the Greeks, about 2,000 years later, who were first intrigued by this relationship.

This triangle also bore an interesting relationship to the Golden Rectangle; that is, if the hypotenuse of the 3-4-5 Triangle were swung in such a way that it formed a right angle with the side measuring three, a rectangle would be formed with a length of five and a width of three. The ratio of 5 to 3 is 1.66666..., a fair approximation of Ø with an error of less than five one-hundredths. So, it appears that, possibly by accident, the rope knotters discovered a triangle that not only contained a right angle but also had harmonic proportions.

Actually, very little is known about how the mathematical discoveries of Ancient Egypt were made. Probably, the common man knew little more than how to make crude calculations that helped him cope with everyday problems. The true keepers of knowledge were the priests, and it is believed that they may well have kept

the more important mathematical finds secret, as was their custom with other scientific knowledge. Some scholars feel that the *Rhind Papyrus*, written by the scribe Ahmes in the 17th century B.C., and believed by some to contain all the mathematical knowledge of the day, was only a rough tool for the common man. Certainly, if that were not the case, some astounding "accidents", as we shall see shortly, occurred during the building of the Great Pyramid of Khufu.

The Babylonians, contemporaries

Below:
Fig. 3.5 The 3-4-5 Triangle expands to approximately a Golden Rectangle.

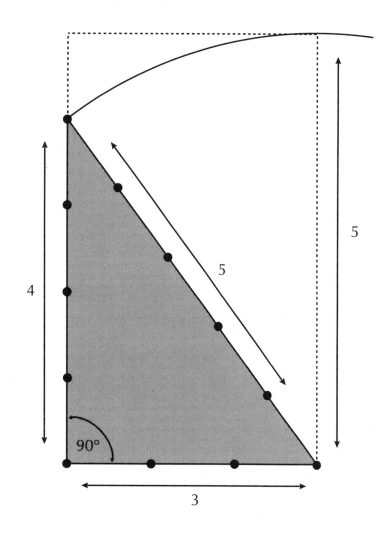

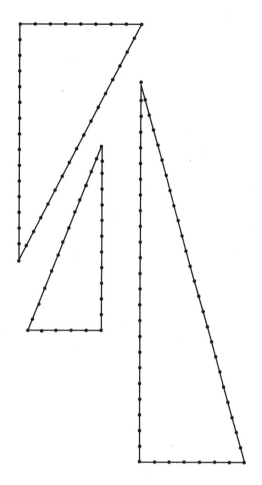

This page:
Fig. 3.6 Three of the fifteen right triangles known to the Babylonians:

Left:
Counting the units:
$8^2 + 15^2 = 17^2$
$64 + 225 = 289$

Center:
$5^2 + 12^2 = 13^2$
$25 + 144 = 169$

Right:
$24^2 + 7^2 = 25^2$
$576 + 49 = 625$

of the Egyptians, discovered fifteen other right triangles whose sides could be measured in whole numbers. These triangles also had the property that the sum of the squares on the legs was equal to the square on the hypotenuse. A few examples are given in Fig. 3.6.

Approximately two thousand years later the Pythagorean Brotherhood of Greece devised a method for finding triangles which had the aforementioned property. Rather, it may be said that the Pythagoreans discovered a rule for finding triples of numbers that could be applied to the sides of right triangles. They were the first to investigate mathematics as a purely abstract endeavor regardless of real world applications. Like philosophy, pure mathematics became a liberal art.

The rule they used follows:

If n is any odd number, then $\dfrac{n^2-1}{2}$ and $\dfrac{n^2+1}{2}$ provide the other two numbers of the triple.

For example: Let n=9.

Then $\dfrac{n^2-1}{2} = \dfrac{9^2-1}{2} = \dfrac{81-1}{2} = \dfrac{80}{2} = 40$ and

$\dfrac{n^2+1}{2} = \dfrac{9^2+1}{2} = \dfrac{81+1}{2} = \dfrac{82}{2} = 41$ and

$9^2 + 40^2 = 41^2$ since $81 + 1600 = 1681$.

This rule did *not* yield every triple of numbers with this special property. For example, $8^2 + 15^2 = 17^2$, but the three numbers 8, 15 and 17 do not fit the rule. Any group of three integers whose squares relate in such a manner is called a **Pythagorean triple** or a **Pythagorean triad**. The Pythagoreans later proved that the sides of *any* right triangle have the property that the square of the hypotenuse is equal to the sum of the squares of the legs. If the legs are labeled a and b, respectively, and the hypotenuse is labeled c, we write $c^2 = a^2 + b^2$. This is called the **Pythagorean Theorem**, $c^2 = a^2 + b^2$, and is proved here.

Remember that the area of a triangle is $\dfrac{1}{2}$ the base times the altitude, and that the area of a square is the square of the side. We can find the area of the small square (c^2), in Fig. 3.7, which is the area of the large square, $(a + b)^2$, minus the area of the four triangles, $4\left(\dfrac{1}{2}ab\right)$.

$c^2 = (a + b)^2 - 4\left(\dfrac{1}{2}ab\right)$

$c^2 = a^2 + 2ab + b^2 - 2ab$

or $c^2 = a^2 + b^2$

This important theorem was a rather distressing revelation for the Pythagoreans who felt that the harmony of the universe was explicable in terms of ratios of whole numbers. The concept of **irrationality** (in numbers) was an anathema to them, but they realized that the ratio of the diagonal of a square to its side could never be expressed by whole numbers no matter how they tried.

$1^2 + 1^2 = 2$ making the square of the hypotenuse 2 and, therefore, the measure of the hypotenuse $\sqrt{2}$ (Fig. 3.8). Since there is no whole number, or fraction, that can be multiplied by itself to produce 2, $\sqrt{2}$ is irrational. Legend has it that this discovery was made during a sea voyage and that the poor fellow who made it was thrown overboard for his efforts.

The discovery of the Pythagorean Theorem altered the course of mathematical thought for the Greeks. From numerical measurement they turned their attention to the geometry of shapes, for what they could not explain with whole number ratios, they could express by geometric figures. We will see further illustrations in Chapter 4.

The 3-4-5 Triangle is probably the most familiar example of the Pythagorean relationship. Two constructions for this special triangle follow.

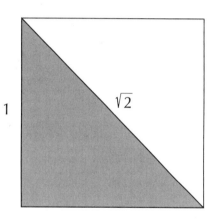

Above:
Fig. 3.7 Illustration of the Pythagorean Theorem.

Left:
Fig. 3.8 The length of the diagonal of a square cannot be expressed in whole number ratios.

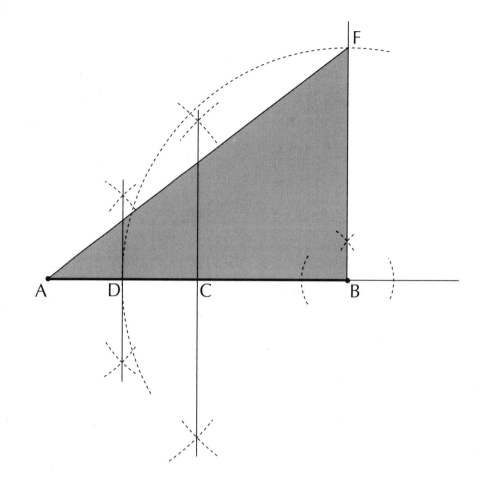

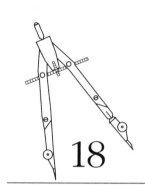

18

Construct a
3-4-5 Triangle
Given the
Longer Leg

Given $\overline{AB}$

1. Bisect $\overline{AB}$ and label its midpoint C. Bisect $\overline{AC}$ and label its midpoint D.

2. Extend $\overrightarrow{AB}$ through B, and construct a perpendicular to $\overleftrightarrow{AB}$ at B.

3. Place the metal tip on B and the pencil tip on D and cut an arc that intersects the perpendicular drawn in the previous step. Label the point of intersection F. Draw $\overline{AF}$.

Now $\triangle ABF$ is a 3-4-5 Triangle with BF:AB:AF = 3:4:5 and $\angle B = 90°$.

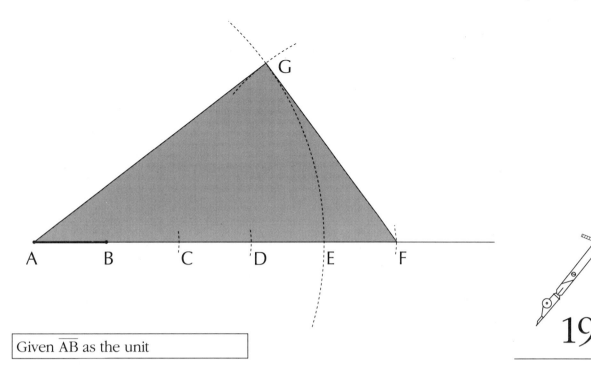

Given $\overline{AB}$ as the unit

1. Extend $\overrightarrow{AB}$ through B. Open the compass so that it measures the same as $\overline{AB}$.

2. Place the metal tip on B and cut an arc that intersects $\overline{AB}$ on the side of B opposite A. Label the point of intersection C.

3. Repeat the previous step three more times, each time placing the metal tip on the most recently constructed point of intersection on $\overrightarrow{AB}$. Label the points of intersection D, E and F respectively.

4. Place the metal tip on A and the pencil tip on E and cut an arc on one side of $\overrightarrow{AB}$. Place the metal tip on F and the pencil tip on C and cut an arc that intersects the one just drawn. Label the point of intersection G.

5. Draw $\overline{AG}$ and $\overline{GF}$.

Now ΔAFG is a 3-4-5 Triangle with FG:GA:AF = 3:4:5 and $\angle G = 90°$.

19

Construct a 3-4-5 Triangle Given the Unit

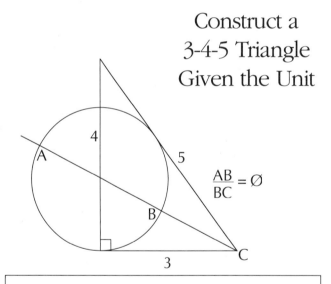

$$\frac{AB}{BC} = \varnothing$$

A connection between the 3-4-5 Right Triangle and $\varnothing$:

Bisect the angle formed by the sides of length 3 and 5 and extend the bisector through the opposite side. The point of intersection of the side and the bisector is the center of a circle that is tangent to the side of length 5 and passes through the vertex of the right angle. The ratio of the diameter (AB) to the other segment of the angle bisector (BC) is $\varnothing$.

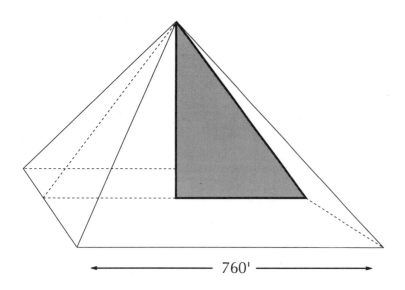

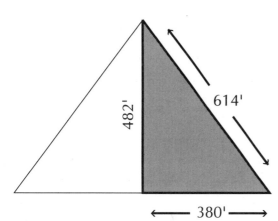

Top:
Fig 3.9 The Triangle of Price in the Pyramid of Khufu

Center and bottom:
Fig 3.10 Harmonic properties of the Triangle of Price. Measurements, rounded to the nearest hundredth, represent $\sqrt{\varnothing}$ and $\varnothing$ to two decimal places.

$$\frac{482}{380} = 1.27$$

$$\frac{614}{482} = 1.27 \qquad \sqrt{\varnothing} \approx 1.27$$

$$\frac{614}{380} = 1.62$$

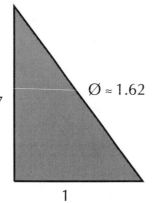

Let us take a backward jump in history once again to the time of the Egyptians and have a closer look at the Great Pyramid of Khufu. As you recall, this pyramid is set on a square base with four triangular faces. A cross section of the pyramid yields a triangle different from the faces and one which has some unique properties.

If an **altitude** is dropped from the **vertex** of the triangle, it bisects the original angle and forms the right triangle that has come to be known as the **Egyptian Triangle** or the **Triangle of Price**, after its discoverer. This triangle has the distinction of being the *only* right triangle whose sides are in **geometric progression.**

Notice the ratios of the lengths of the three sides of the Triangle of Price in the Pyramid of Khufu (Fig. 3.9). Incredibly enough, $1.27 = \sqrt{\varnothing}$ to two decimal places. Therefore, if we relabel the sides of this triangle (Fig 3.10), naming the shortest side "one", or unity, the remaining sides are $\sqrt{\varnothing}$ and $\varnothing$. The Triangle of Price, then, like the pentagram and pentagon mentioned in Chapter 2, contains harmonic properties. Construction 20 is based on the fact that the ratio of a diagonal to a side of a regular pentagon is $\varnothing$. Construction 21 demonstrates how nearly alike are the 3-4-5 Triangle and the Triangle of Price. You will recall that the 3-4-5 Triangle can be expanded into a rectangle which very closely approximates a Golden Rectangle. It is the Triangle of Price which expands precisely to a Golden Rectangle. Because they are so similar (see Fig. 3.13 on page 93), the 3-4-5 Triangle, the Triangle of Price and the **Harmonic Triangle**, discussed in the next section, may be used interchangeably for most design purposes.

The Pythagoreans were struck by the fact that phenomena which are physically most diverse exhibit identical mathematical properties.

Morris Kline
Mathematics in Western Culture

This page:
Joan Coronis. Student Work.
Chapter 3 Project 3. Watercolor on paper with dimensional cut paper elements.

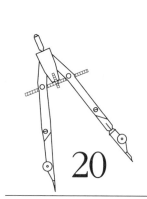

20

Construct a
Triangle of
Price Given a
Regular
Pentagon

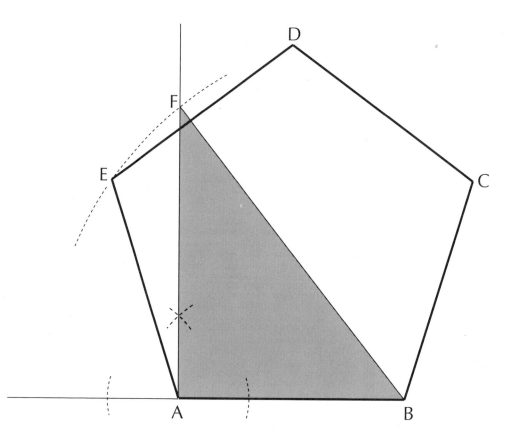

Given regular pentagon ABCDE

1. Extend $\overleftrightarrow{AB}$ through A.

2. Construct a perpendicular to $\overleftrightarrow{AB}$ through A.

3. Place the metal tip on B and the pencil tip on E and cut an arc that intersects the perpendicular in the previous step. Label the point of intersection F. Draw $\overline{BF}$.

Now ΔABF is a Triangle of Price with AB = 1, AF = $\sqrt{\varnothing}$ and BF = $\varnothing$.

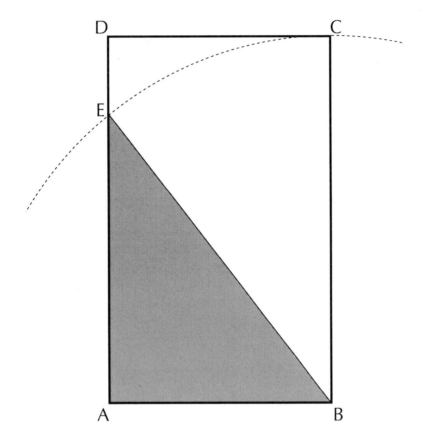

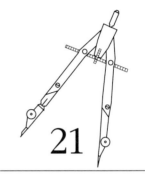

21

Construct a
Triangle of
Price Given a
Golden
Rectangle

Given Golden Rectangle ABCD

---

1. Place the metal tip on B and the
pencil tip on C and cut an arc that
intersects $\overline{AD}$. Label the point of intersection E.

2. Draw $\overline{EB}$.

Now ΔABE is a Triangle of Price with
AB = 1, AE = $\sqrt{\emptyset}$ and EB = ∅.

As we explained earlier on, standardized measurement became a necessity for the Egyptians. In earlier times, when building only for themselves, men could use the lengths of particular parts of their own bodies as constant units for measure. We still refer to the height of a horse in terms of *hands*, or say an object is so many *heads* high, and use rulers that are a *foot* long.

The Egyptians used the foot, palm, fingers, and forearm for measuring. Fig. 3.11 shows the relationship between digits, palms, and **cubits**. Actually, when standardized, the Egyptians used two variations for the cubit; the **little cubit** which measured just under half a meter and the **Royal Cubit** which measured just over half a meter. See Appendix B for the relationship between the little cubit, the Royal Cubit and Ø.

If a right triangle is formed with a hypotenuse that measures a Royal Cubit and one leg that measures a little cubit, the result is a triangle called the Harmonic Triangle. Oddly enough, the remaining side has a measure that is an average obtained from the various standard *foot* measurements used among different cultures of that time. Geometrically, this triangle is the one formed by the radius and **apothem** of a regular pentagon, and a method for constructing it follows. The subsequent construction shows the relationship between the Harmonic Triangle and an inscribed pentagon.

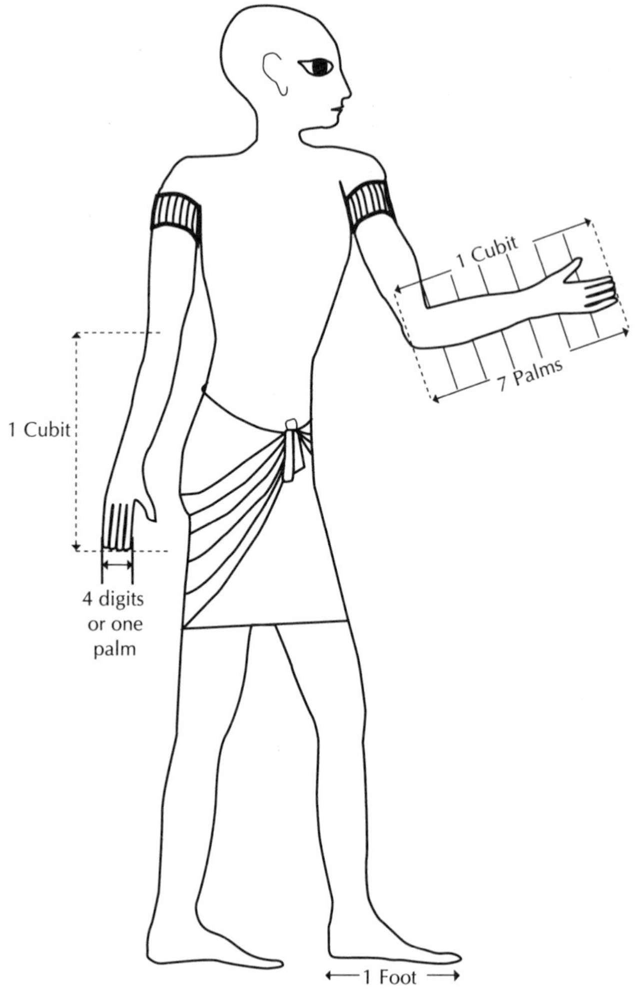

Above:
Fig. 3.11 Elements of Egyptian measurements.

Below:
Fig. 3.12 The Harmonic Triangle.

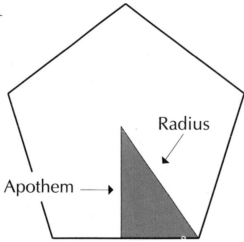

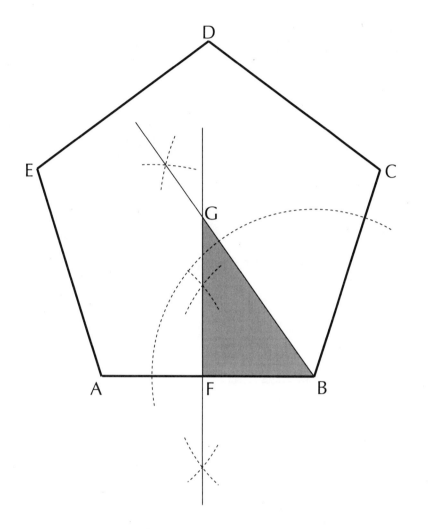

22

Construct an Harmonic Triangle Given a Regular Pentagon

Given regular pentagon ABCDE

1. Construct the perpendicular bisector of $\overline{AB}$. Label the point of intersection F. (F is the midpoint of $\overline{AB}$.)

2. Bisect ∠B and extend it so that it intersects the perpendicular drawn in the previous step. (Refer to Construction 3). Label the point of intersection G.

Now ΔFBG is an Harmonic Triangle.

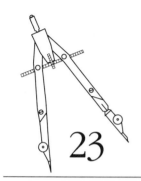

23

Construct an
Harmonic
Triangle Given
a Regular
Pentagon
Inscribed in a
Circle

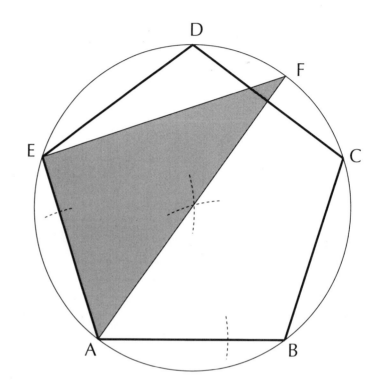

Given regular pentagon ABCDE
inscribed in a circle

1. Bisect ∠A and extend the bisector
to intersect the circle. Label the point
of intersection F. (AF is a **diameter** of
the circle.)

2. Draw $\overline{EF}$.

Now △AEF is an Harmonic Triangle.

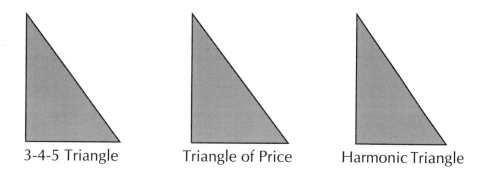

3-4-5 Triangle   Triangle of Price   Harmonic Triangle

Detail: comparison of vertices when bases are aligned

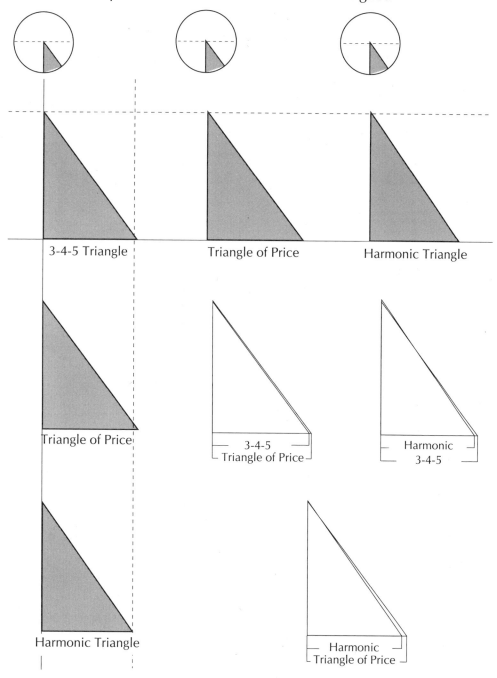

This page:
Fig. 3.13 Comparison of the 3-4-5 Triangle, Triangle of Price and Harmonic Triangle.

You will notice minute discrepancies among the three triangles both in the length and the width. The differences are not sloppiness of computer diagram construction, but subtleties of variation in the legs and hypotenuse of each triangle. They could be considered interchangeable for design purposes if used in a small size. The differences would be magnified if the sizes were to increase drastically. In the natural world, however, minute changes, such as these, at the beginning of a process could have grave consequences long range. Therefore, we encourage you to be sensitive to such details. As one artist has suggested, "God is found in the details."

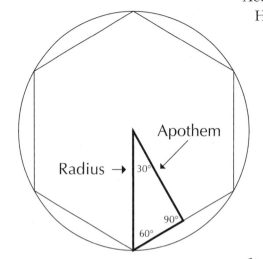

According to some sources, the Harmonic Triangle was used as the standard drafting tool of the Middle Ages. If so, it has since been replaced by another right triangle that does not have harmonic proportions, but whose sides are in a much simpler ratio. This is the one called the **30-60-90** or **Timaeus Triangle**. It is the one built from the radius and apothem of the regular **hexagon,** rather than the pentagon, and its angles measure 30°, 60°, and 90°, respectively.

If we let the side opposite the 30° angle measure 1, then the hypotenuse measures 2 and, by the Pythagorean Theorem, the remaining side measures $\sqrt{3}$. Similarly, if b is any number, say x, c is twice its length or 2x, and a is $x\sqrt{3}$, since $a^2 + b^2 = c^2$.

Some scholars feel that with the change from the use of the Harmonic to the 30-60-90 Triangle much of the beauty and harmony in architectural design was lost. The Harmonic Triangle is based on the pentagon found in many of the organic forms in nature, but never in static or crystalline forms. The 30-60-90 Triangle, on the other hand, comes from the hexagon which is found in some lower order creatures but which abounds in inorganic forms in Nature. To some, this relationship with the non-living suggests sterility, hence lack of good proportion, and yet Plato felt this was the most beautiful of all triangles. We feel that the crystalline and the organic are two aspects of the unity that is Nature. Moral judgments of better or worse do not apply.

This page:
Fig 3.14 Top to bottom: Relationship of hexagon to 30-60-90 Triangle.

30-60-90 Triangle.
$a^2 + b^2 = c^2$
$a^2 + 1^2 = 2^2$
$a^2 + 1 = 4$
$a^2 = 3$
$a = \sqrt{3}$

30-60-90 Triangle.
$a^2 + x^2 = (2x)^2$
$a^2 + x^2 = 4x^2$
$a^2 = 3x^2$
$a = \sqrt{3x^2}$
$a = x\sqrt{3}$

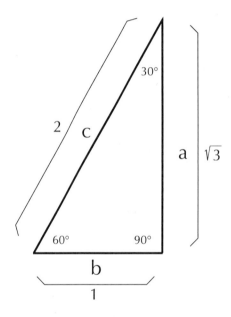

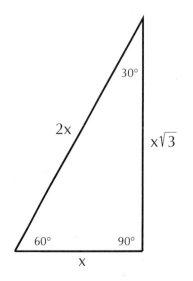

The other standard draftsman's triangle is the one that is derived by dividing a square along its diagonal. This triangle, then, has a pair of congruent sides and a pair of congruent angles, the latter always measuring 45°.

We have already seen that if the sides of the square measure 1, then the diagonal measures that irrational number $\sqrt{2}$ that caused the Pythagoreans such discomfort. Yet, it is the Pythagorean Theorem which allows us to find the measure of the hypotenuse regardless of the measure of the side.

We know that $a^2 + b^2 = c^2$, but this time $a = b$, so we have

$a^2 + a^2 = c^2$

$2a^2 = c^2$

$\sqrt{2a^2} = c$

$a\sqrt{2} = c$

So if $a = 1$, $c = \sqrt{2}$

if $a = 2$, $c = 2\sqrt{2}$

if $a = 3$, $c = 3\sqrt{2}$,

and so on. This triangle is called the **45-45-90 Triangle**, or the **Isosceles Right Triangle**, and its construction follows on the next page.

Any triangle that has two congruent sides is called **isosceles,** and any isosceles triangle has a pair of congruent angles called **base angles**. The third angle is called the **vertex angle.**

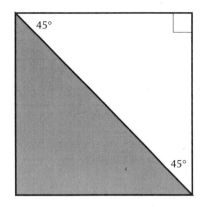

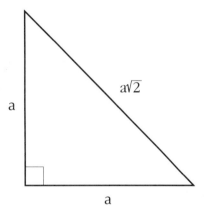

Top:
Fig. 3.15 Isosceles Right Triangle and square.

Left:
Fig. 3.16 Isosceles Right Triangle.

Bottom:
Fig. 3.17 Isosceles triangles.

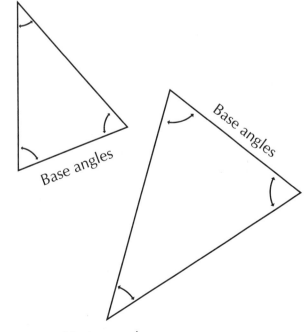

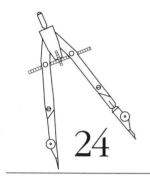

24

Construct an Isosceles Triangle Given the Base

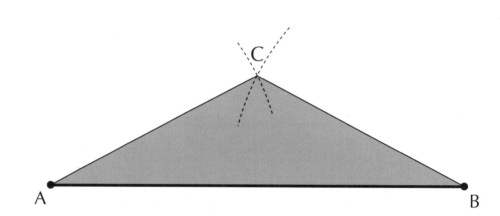

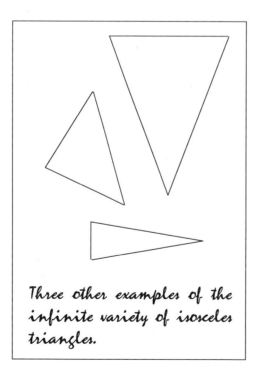

Three other examples of the infinite variety of isosceles triangles.

Given $\overline{AB}$, the **base**

1. Open the compass so that it measures the desired length of the **legs** of the triangle. *(Note: this measure must be greater than half of AB or a triangle cannot be formed.)* Place the metal tip on A and cut an arc above $\overline{AB}$. Without changing the setting, place the metal tip on B and cut an arc that intersects the one previously drawn. Label the point of intersection C.

2. Draw $\overline{AC}$ and $\overline{BC}$.

Now $\triangle ABC$ is isosceles with $\overline{AC} \cong \overline{BC}$ and $\angle A \cong \angle B$.

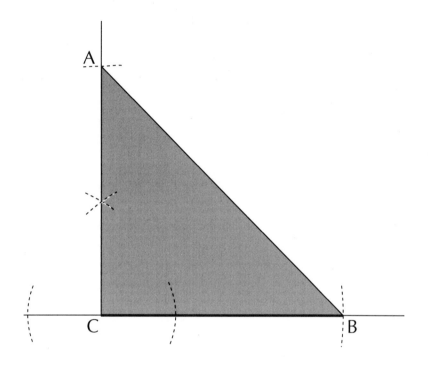

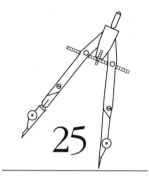

25

Construct an Isosceles Right Triangle Given a Leg

Given $\overline{CB}$

1. Extend the line through C and construct a perpendicular to $\overline{CB}$ through C.

2. Open the compass to measure CB and, with the metal tip on C, cut an arc that intersects the perpendicular just constructed.

3. Label the point of intersection A, and draw $\overline{AB}$.

Now $\triangle ABC$ is an Isosceles Right Triangle with $\overline{CA} \cong \overline{CB}$. Also $\angle A \cong \angle B$ and each measures 45°.

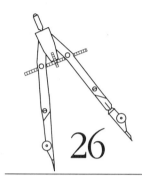

26

Construct a Right Triangle in a Semicircle Given the Circle

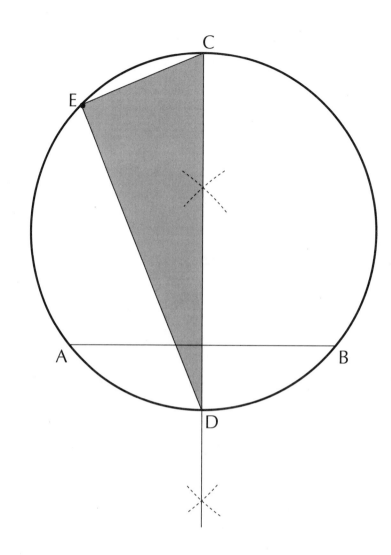

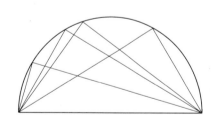

Note: any triangle inscribed in a semicircle is a right triangle. Constructions 26 and 27 provide methods whereby you can obtain right triangles by first constructing a semicircle.

Given a circle

1. Draw a **chord** and label its endpoints A and B.

2. Construct the perpendicular bisector of $\overline{AB}$ and extend it through the circle. Label the points where it intersects the circle C and D. ($\overline{CD}$ is a **diameter** of the circle.)

3. Anywhere on the circle, choose point E. Draw $\overline{EC}$ and $\overline{ED}$.

Now ΔCED is a right triangle with ∠E the right angle.

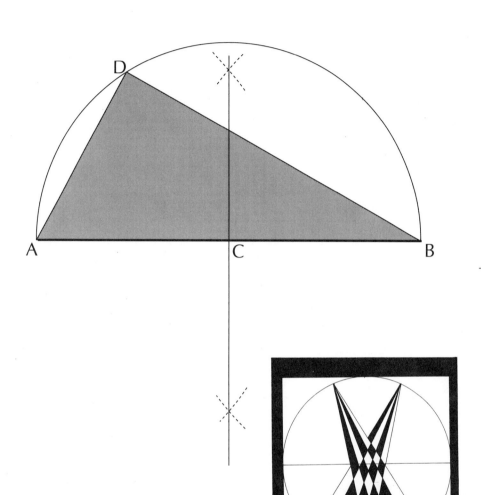

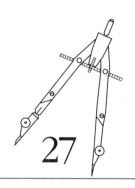

27

Construct a Right Triangle in a Semicircle Given the Diameter

Given $\overline{AB}$

1. Bisect $\overline{AB}$ and label its midpoint C.

2. Place the metal tip on C and cut an arc from A to B.

3. Choose any point D on arc AB, and draw $\overline{DA}$ and $\overline{DB}$.

Now $\triangle ADB$ is a right triangle with $\angle D$ the right angle.

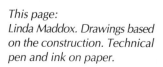

This page:
Linda Maddox. Drawings based on the construction. Technical pen and ink on paper.

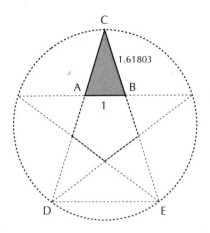

1.61803

1

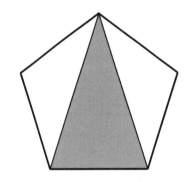

Top and right:
Fig. 3.18 Triangle of the Pentalpha or Golden Triangle.

Bottom:
Linda Maddox. These drawings further demonstrate the use of the pentagram and Golden Triangle as elements of design in dividing areas harmoniously. Technical pen and ink on paper.

One isosceles triangle relates directly to the pentagon and its pentagram, and picks up the harmonic proportions inherent in these figures. In Fig. 3.18 ΔABC is isosceles with base angles that measure 72° and a vertex angle of 36°. As you recall from Chapter 2, $\frac{CB}{BA} = \emptyset \approx 1.61803$. If we draw $\overline{DE}$, we have another triangle, ΔDEC, which is **similar** to ΔABC. Since similar triangles have corresponding sides in proportion, we have $\frac{CB}{BA} = \frac{CE}{ED} = \emptyset$. This triangle was called the **Sublime Triangle**, or the **Triangle of the Pentalpha**, and Fig. 3.18 demonstrates that it is a simple matter to construct it by drawing the two diagonals from one vertex of a regular pentagon. The following construction shows the connection between the Triangle of the Pentalpha and the Golden Rectangle. Because of this relationship it has come to be known as the **Golden Triangle**, which is the name we shall use from this point on.

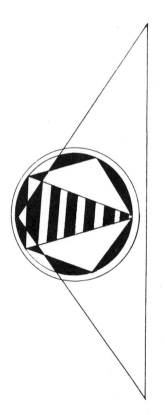

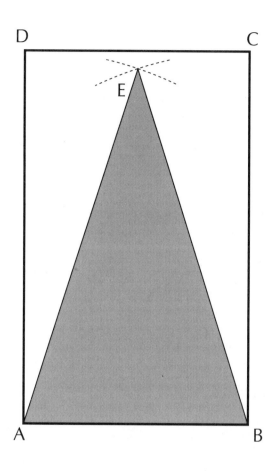

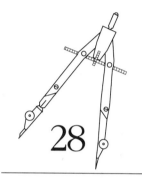

28

Construct a Golden Triangle Given a Golden Rectangle

Given Golden Rectangle ABCD

1. Place the metal tip on B and the pencil tip on C and cut an arc in the interior of ABCD.

2. Without changing the setting, place the metal tip on A and cut an arc that intersects the one drawn in Step 1.

3. Label the point of intersection E, and draw $\overline{EA}$ and $\overline{EB}$.

Now ΔABE is a Golden Triangle with $\dfrac{EB}{AB} = \dfrac{EA}{AB} = \varnothing$.

Left:
Linda Maddox. Drawing based on the Golden Triangle. Technical pen and ink on paper.

This page:
*Fig. 3.19 There is an intimate relationship between the pentagon and the Golden Triangle in a figure that is known as the **Lute of Pythagoras**. In the construction of this figure, a "ladder" is built within a Golden Triangle. Construction 29 , the one that follows, gives the details of the figure.*

Technical pen drawings by Linda Maddox further demonstrate the use of the Lute of Pythagoras as a design element.

Given Golden Triangle OPQ

1. Open the compass to measure $\overline{PQ}$. Without changing the setting, place the metal tip first on P and then on Q, cutting arcs on the opposite side of the triangle. Label the points of intersection R and S. Continue this process upward through the triangle, each time, setting the compass by the previously cut pair of arcs (eg. open the compass to measure RS to cut the arcs that intersect at the points you will label T and U). Join the points of intersection by line segments.

2. Form the pentagonal figures within the triangle by joining corresponding points on either side of the triangle to the point where the intersecting segments meet in the previous "rung" (eg. join T and U to M, respectively).

3. Point Z is determined by setting the compass to measure QS and then cutting arcs by placing the metal tip on P and Q, respectively.

4. Draw $\overline{PZ}$, $\overline{ZQ}$, $\overline{RZ}$, $\overline{SZ}$. Note that RSZ is an inverted Golden Triangle.

5. The small interior pentagram is constructed by connecting non-consecutive vertices of the pentagon formed by the diagonals of RPZQS.

Now the figure is a Lute of Pythagoras. Not only does this figure contain all the already mentioned Phi proportions found in the pentagon, pentagram, and Golden Triangle, but the rungs of the ladder also divide the sides of the larger Golden Triangle into Phi proportions. That is: $\dfrac{QS}{SU} = \dfrac{SU}{UW} = \dfrac{UW}{WY} = \emptyset$.

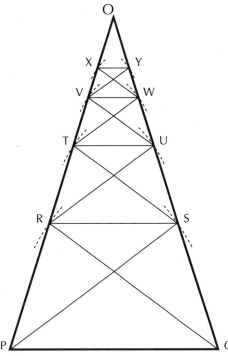

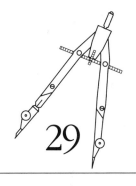

29

Construct a
Lute of
Pythagoras
Given a
Golden
Triangle

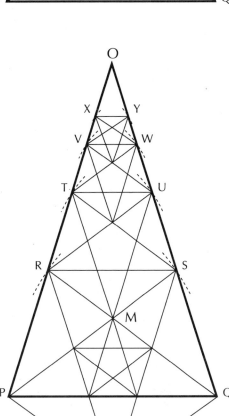

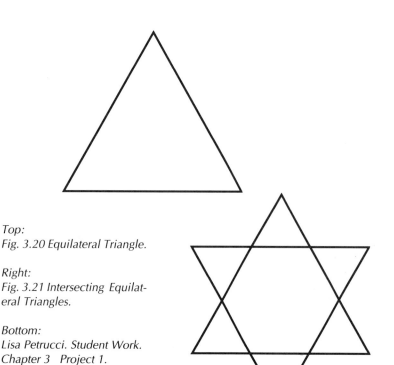

Top:
Fig. 3.20 Equilateral Triangle.

Right:
Fig. 3.21 Intersecting Equilateral Triangles.

Bottom:
Lisa Petrucci. Student Work. Chapter 3   Project 1.

The last triangle we wish to consider is the **Equilateral Triangle**. It is the one whose sides and angles are all congruent. The sides may measure any length, but the angles will always be 60°.

The equilateral triangle has been used in the design of symbols throughout history. It is the Christian symbol for the Trinity while both the Ancient Chinese and Japanese considered it the symbol for Heaven, Earth and Man.

When known as the Jewish Star of David, the intersecting pair of Equilateral Triangles symbolizes the power of Jehovah moving outward from the center as well as being seen as the sixfold symbol of creation and perfection. In India the interlocking of the two triangles represents Shiva and Shaki forming the Wheel of Vishnu. For the ancient Egyptians, the symbol stood for fertility by suggesting the sexual union of the pagan goddess Ashtaroth with the god Adonis. As Solomon's Seal, it symbolizes the synthesis of the four elements of earth, air, fire and water. It also acts as a symbol for the union of complements: the triangle pointing upwards represents Man striving for Heaven while the downward pointing triangle represents Man struggling with earthly concerns.

If the equilateral triangle also appeals to you as a design element, its construction is a very simple matter.

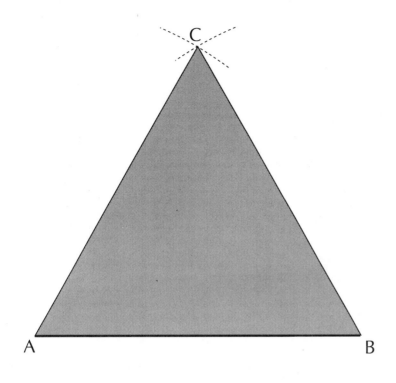

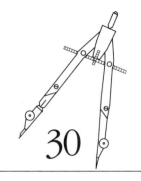

30

Construct an Equilateral Triangle Given a Side

Given $\overline{AB}$

1. Place the metal tip on A and the pencil tip on B and, without changing the setting, cut an arc on one side of $\overleftrightarrow{AB}$.

2. Without changing the setting, place the metal tip on B and cut an arc that intersects the one previously drawn. Label the point of intersection C.

3. Draw $\overline{AC}$ and $\overline{BC}$.

Now ΔABC is equilateral with $\overline{AB} \cong \overline{BC} \cong \overline{AC}$. Also, ∠A ≅ ∠B ≅ ∠C, and each measures 60°.

Hint: To construct an equilateral triangle given a circle, you must first construct a hexagon. This is gotten by setting the compass at the radius (which measures half the diameter) and cutting arcs around the circle. For the triangle, connect every other vertex.

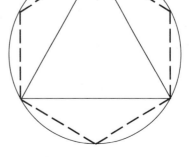

Facing pages:
Implied equilateral triangles
found in forms in the natural
world.

106

Fig. 3.22 illustrates the harmonic subdivision of the equilateral triangle. Each of the sides of ΔABC has been divided into Ø proportions by finding the Golden Cut (see Construction 10). That is: $\frac{DB}{DA} = \frac{AE}{EB} = \varnothing$. The same relationship holds for the other two sides.

Let us briefly examine the relationship between the equilateral triangle and the 30-60-90 Triangle. Since the equilateral one has angles of 60°, we need only bisect one of these to divide the angle into two 30° angles, and the triangle into two congruent 30-60-90 Triangles. Because of the symmetric properties of the equilateral triangle, any angle bisector is also the perpendicular bisector of the opposite side, and provides the requisite right angle for the 30-60-90 Triangle.

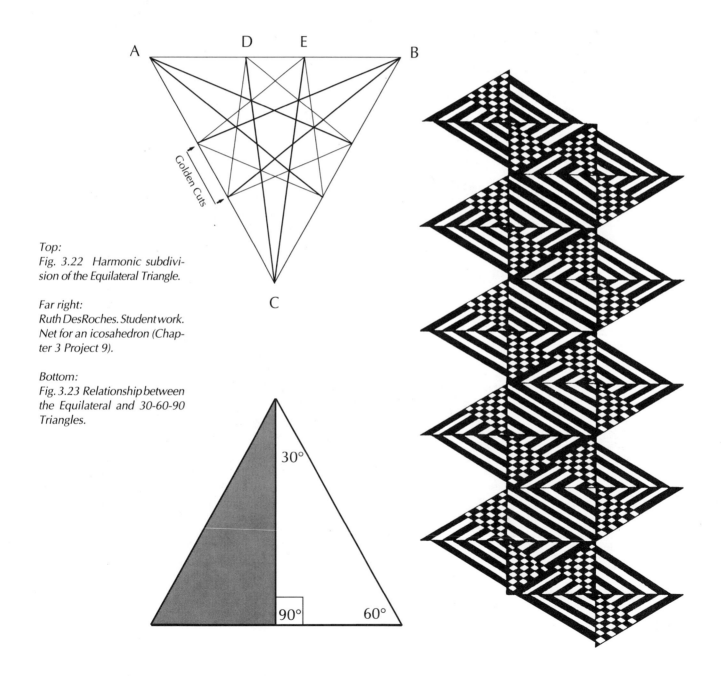

Top:
Fig. 3.22 Harmonic subdivision of the Equilateral Triangle.

Far right:
Ruth DesRoches. Student work. Net for an icosahedron (Chapter 3 Project 9).

Bottom:
Fig. 3.23 Relationship between the Equilateral and 30-60-90 Triangles.

Special Triangles

Name(s)	Type	Angles	Sides	Diagram
3-4-5 Triangle Rope Stretcher's Rope Knotter's Pythagorean	Right Triangle	36°52'12" 53°7'48" 90°	*Only* right triangle whose sides are in arithmetic progression, in the ratio 3:4:5.	
Triangle of Price Egyptian Triangle	Right Triangle	38°30', 51°30', 90°	*Only* right triangle whose sides are in geometric progression, in the ratio of $1:\sqrt{Ø}:Ø$.	
Harmonic Triangle	Right Triangle	36°, 54°, 90°	In the ratio of 1 foot: 1 little cubit: 1 Royal Cubit.	

The above three triangles may be interchanged for most design purposes.

Name(s)	Type	Angles	Sides	Diagram
30-60-90 Triangle Timaeus Triangle	Right Triangle	30°, 60°, 90°	In the ratio of $1:\sqrt{3}:2$.	
Isosceles Right Triangle 45-45-90 Triangle	Right Triangle Isosceles Triangle	45°, 45°, 90°	In the ratio of $1:1:\sqrt{2}$.	
Golden Triangle Sublime Triangle	Isosceles Triangle	36°, 72°, 72°	Two congruent sides. Ratio of leg to base is Ø.	
Equilateral Triangle	Equilateral Triangle	60°, 60°, 60°	Three congruent sides.	

109

Problems

[1] Each of the symbols above in some way utilizes the concept of equilateral triangles. Reconstruct each one on a separate sheet of paper. For ease of construction, let the side of the largest triangle in each symbol measure a minimum of four inches.

[2] Construct one of each of the Special Triangles that appear on the chart on the previous page.

[3] Construct a Golden Triangle with leg length of at least five inches. Construct the Lute of Pythagoras within the triangle (see Construction 29).

[4] Construct an equilateral triangle with sides at least four inches long. Subdivide the sides harmonically and join the points as in the illustration in Fig. 3.22. Strengthen the design by alternately coloring in areas with black.

[5] Go out into the landscape with a friend(s), some pegs and some cord. Using these materials, mark off an area of ground. Now, using the process of triangulation, divide the area into triangles, measure the individual ones and add together to obtain the total of the bounded area. Document this experience with photographs, drawings or diagrams.

6 Given the illustrations of butterflies and moths, escribe a triangle around each. The two already done in dotted lines illustrate the method. Be sure the triangles are tangent to the wings in as many places as possible. What kinds of special triangles occur? You will need to use a **protractor** to measure the angles of the triangles. Summarize or chart your findings.

Projects

1 Now that you have learned to construct several different types of triangles, create an interesting pattern, which may be either two- or three-dimensional, using:

   a. the same shape triangle, but altering the sizes and using a **monochromatic** coloring scheme,

   b. two different types of triangles and a **complementary** color scheme, *or*

   c. a variety of triangular shapes and an **analogous** color scheme.

2 Using any materials whatsoever, construct a pyramid that is at least 12" high, on a square base. (This is more complicated than it appears.) Try to make it aesthetically pleasing as well as accurate. Think of what the Egyptians accomplished.

3 Given the information in Problem 6, create two imaginary butterflies that look different but will fit into the same (any one of your choice) Special Triangle. Use whatever materials catch your fancy.

4 On illustration board, bristol or watercolor paper, construct a Golden Rectangle in dimensions of your own choosing. Using the process of triangulation, divide the area enclosed by the rectangle. Color by using *one* of the following:

   a. Black and one other color

   b. Black and white

   c. Five values of gray

   d. Five variations on a single hue

   e. A pair of complementary colors in light and dark variations

   f. A triad of colors.

Above:
Julie Conway. Student Work.
Chapter 3   Project 3. Ink on paper.

5 Now that you have learned of other useful triangles, construct a set of heavy cardboard or plastic templates of the triangles that you would find useful. Make them attractive, functional and legible.

6 Assume you have your own business. Use the harmonic subdivisions of an equilateral triangle to design a logo for your company. The choice of black and white, value, or color is yours, but the finished work should be done on illustration board and the triangle should have sides that measure a minimum of ten inches. Use acrylic, casein, watercolor, inks, or gouache to obtain good surface consistency.

7 Make a kite using the Lute of Pythagoras as a pattern. Construct the Lute out of tissue paper, acetate, garbage bags, cellophane or fabric in any color scheme of your choice. Support your kite with balsa wood strips and string around the edges. Add a tail and make a windy day magical. (This is a very time consuming project despite its apparent simplicity.)

8 Using the photo of the cat's face at the beginning of the chapter, isolate the triangular areas and create a stylized, more geometric cat face.

9 Sculptural Form:
Twenty equilateral triangles join to form a figure called the icosahedron. The pattern below shows how to lay out the 20 triangles in the plane. By bending, folding, scoring, and pulling together, the icosahedron is formed. Create your own variation on this Platonic Solid by using one of the following:

a. Make a template out of heavy paper or acetate following the directions for Construction 30. Be sure to include glue flaps. Use this template to construct all 20 units.

b. Divide the triangle itself into harmonic divisions following the directions in Fig. 3.22.

c. Join all 20 units, or multiples of 20, to create a sculpture.

d. Use materials of your choice to make the finished form. Consider the possibilities of using:
   1. felt material
   2. patchwork units
   3. knitted units
   4. crocheted units
   (These would have to be backed or stuffed in some manner to add rigidity.)
   5. Fome-Cor board
   6. Plexiglas
   7. sheet copper, brass or aluminum
   8. colored acetates
   9. stained glass pieces
   10. clay units
   11. any other materials that you might fancy.

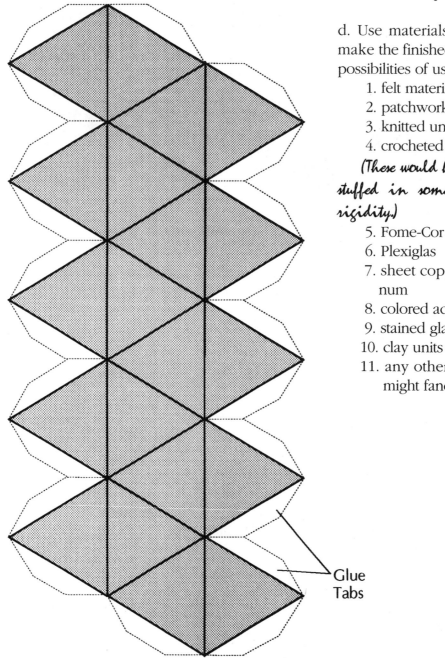

Glue
Tabs

Further Reading

Eves, Howard. *An Introduction to the History of Mathematics.* 3rd Ed. New York: Holt, Rinehart and Winston, 1969.

Fleming, William. *Arts and Ideas.* New York: Holt, Rinehart and Winston, 1980.

Gillings, Richard J. *Mathematics in the Time of the Pharaohs.* New York: Dover Publications, Inc., 1972.

Grillo, Paul Jacques. *Form, Function and Design.* New York: Dover Publications, Inc., 1960.

Hogben, Lancelot. *Mathematics in the Making.* Garden City, New York: Doubleday and Company, Inc., 1960.

Mendelssohn, Kurt FRS. *The Riddle of the Pyramids.* London: Thames and Hudson, 1974.

Westwood, Jennifer, Editor. *The Atlas of Mysterious Places.* New York: Weidenfeld and Nicolson, 1987.

4 Dynamic Rectangles

Composition is to the artist what formulas are to the mathematician. Each is a skeletal understructure that gives support to the flesh of an idea. For the visual artist, the idea of composition is rooted in the concept of spatial division. In drawing, painting or design, this involves bounding a portion of the plane and then subdividing the enclosed area into meaningful divisions which are then used with either non-objective or representational images.

Prior to the time of the Ancient Egyptians, all spatial directions were given equal consideration. For the Egyptians, the rectangle became the supreme compositional format, and the relationship between the vertical and the horizontal at a 90° angle was embedded in human consciousness.

The first concern of the artist is, "What size shall this rectangle be?" When painting was more closely allied with the mural form, it was the intended architectural space-surface that determined this. With the advent of easel

painting, the artist was left to his or her own fancies as to what would be the size of this portable rectangle. Naturally, if the works were meant for the palaces of kings or the parlors of burghers, the ultimate sizes of the finished works would be influenced by this factor. But, if the artist were no longer tied to a particular audience or place, then the rectangle would be an arbitrary choice. There should always be a connection between the scale of an idea and its particular presentation, but within these confines, how does an artist go about choosing the "right" ratio between the vertical and the horizontal? And after that is determined, the artist must divide the areas within the

Above:
Comparison of directional emphasis in cave paintings and Egyptian wall relief.

117

Below:
Fig. 4.1 Rectangles in which lengths of sides are in whole number ratios.

rectangle so as to provide both structure and meaning. For space, beyond its physical limitations, acts symbolically as well.

The artist could choose to use rectangles whose sides are in whole number ratios as seen in the examples in Fig. 4.1. Although, individually, each rectangle can be evenly subdivided into square units, when used in combination a composition may lack harmony between the parts and the whole. The unit length may vary, as illustrated, and is dictated by the particular instrument of measure. For example, a rectangle may measure $3\frac{7}{8}$" by $4\frac{1}{2}$" and still its sides are in whole number ratio if the unit square measures 1/8 inch on a side.

There is, however, a group of rectangles, all derived from the square, which gives us a family that relates proportionally and provides harmony. Members of this group are called the **Dynamic Rectangles**, or sometimes the **Euclidean Series**. They afford harmony within each rectangle and relate harmoniously to each other. As we shall see when we examine the structure of the rectangles, there are subgroups of the family that work well together. Since the square is an integral part of each member of the set, it can be used harmoniously with any of these rectangles within a given composition.

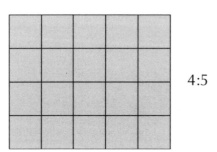

4:3

4:5

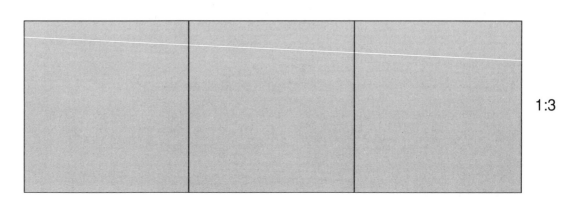

1:3

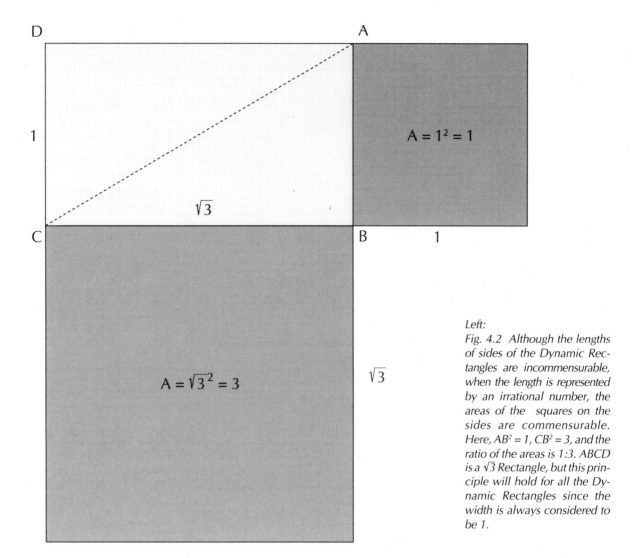

D · A

1

$\sqrt{3}$

$A = 1^2 = 1$

C · B · 1

$A = \sqrt{3}^2 = 3$

$\sqrt{3}$

Left:
Fig. 4.2 Although the lengths of sides of the Dynamic Rectangles are incommensurable, when the length is represented by an irrational number, the areas of the squares on the sides are commensurable. Here, $AB^2 = 1$, $CB^2 = 3$, and the ratio of the areas is 1:3. ABCD is a $\sqrt{3}$ Rectangle, but this principle will hold for all the Dynamic Rectangles since the width is always considered to be 1.

Dynamic Rectangles mirror some of the qualities of life. The subdivision of the parent rectangle produces proportionately related units which allow for continued similarity of shape which is suggestive of the principle of growth found in Nature. The rectangles provide structural armatures which are essential for any good composition and design.

Except in certain cases, these rectangles do not have sides in whole number ratios. One side serves as the unit measure and the other side is named by the square root of one of the natural numbers other than 1, so that the sides of the Dynamic Rectangles are in such ratios as $1:\sqrt{2}$; $1:\sqrt{3}$; $1:\sqrt{4}$, etc. Except in the cases where the **radicand** is a perfect square, i.e., 4, 9, 16, etc., the length of the rectangle indicates an irrational number. Because of the appearance of these irrational numbers, the Dynamic Rectangles are sometimes called **root rectangles**. A Root-two Rectangle is the one whose sides measure 1 and $\sqrt{2}$.

In both the construction of the Dynamic Rectangles and their subsequent subdivisions, the diagonal plays an important role. It is used repeatedly in the following construction for setting the compass, although it is never actually drawn.

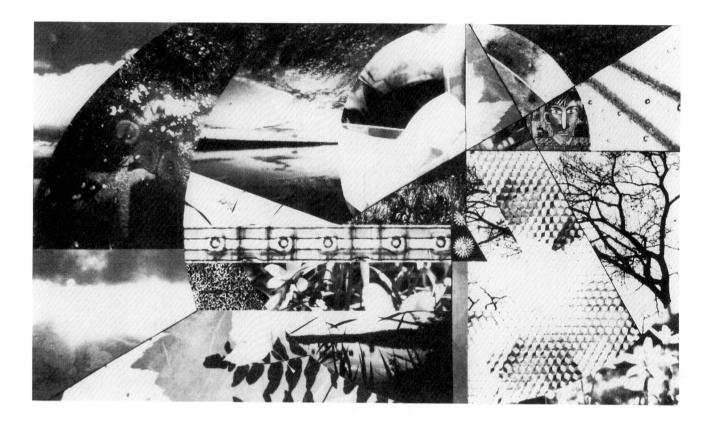

Above:
Priscilla Dullea. Student Work. Variation on Project 3 Chapter 4. Collage using cut paper magazine images.

The Unit Square

When we speak of the **unit square**, we mean the square whose side will serve as the unit measure (or the number 1) for all measurements taken subsequently within a given situation. This unit does not stand for a particular length within any system of measure. Rather, once the unit is determined, it becomes the basis for all measurements that follow. As the choice of the unit is arbitrary, determined by specific demands, it may vary from one situation to another. For example, the unit used to measure the length of a field will differ drastically from the one used to measure a section of galactic space or the diameter of a single cell. Construction 15 in Chapter 2 provides the method for constructing a square. The size of the unit is left to the individual.

In design, the square is often perceived as a static form with few dynamic properties. This feeling may result from not fully exploring its potential; relying primarily on midpoints for structural armatures. Designers would be better served by a more comfortable relationship with geometry. Construction 31 shows how to use the square to develop the Dynamic Rectangles and Construction 32 gives one example of a way to investigate some of its dynamic qualities.

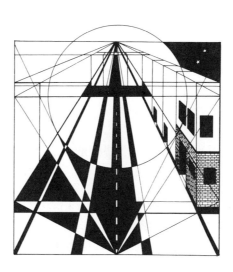

Top:
David Cockroft. Student Work. Chapter 4   Project 1. Ink and metallic markers on paper.

Bottom:
Linda Maddox. Dynamic qualities of the unit square. Technical pen and ink on paper.

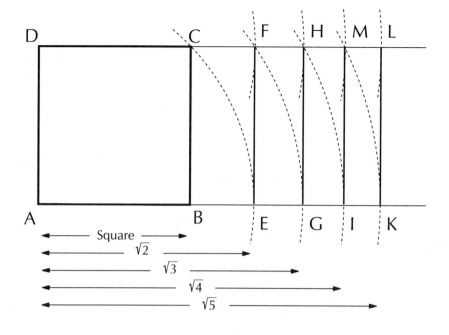

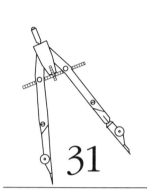

31

Generate the Dynamic Rectangles from the Unit Square

Given square ABCD

---

1. Extend $\overrightarrow{AB}$ and $\overrightarrow{DC}$.

2. Place the metal tip on A and the pencil tip on C and cut an arc that intersects $\overrightarrow{AB}$. Label the point of intersection E.

3. Without changing the compass setting, place the metal tip on D and cut an arc that intersects $\overrightarrow{DC}$. Label the point of intersection F, and draw $\overline{EF}$.

Now AEFD is a $\sqrt{2}$ Rectangle.

4. Place the metal tip on A and the pencil tip on F and cut an arc that intersects $\overrightarrow{AB}$. Label the point of intersection G.

5. Without changing the compass setting, place the metal tip on D and cut an arc that intersects $\overrightarrow{DC}$. Label the point of intersection H, and draw $\overline{GH}$.

Now AGHD is a $\sqrt{3}$ Rectangle.

6. Place the metal tip on A and the pencil tip on H and cut an arc that intersects $\overrightarrow{AB}$. Label the point of intersection I.

7. Without changing the compass setting, place the metal tip on D and cut an arc that intersects $\overrightarrow{DC}$. Label the point of intersection M, and draw $\overline{IM}$.

Now AIMD is a $\sqrt{4}$ Rectangle.

8. Place the metal tip on A and the pencil tip on M and cut an arc that intersects $\overrightarrow{AB}$. Label the point of intersection K.

9. Without changing the compass setting, place the metal tip on D and cut an arc that intersects $\overrightarrow{DC}$. Label the point of intersection L, and draw $\overline{KL}$.

Now AKLD is a $\sqrt{5}$ Rectangle.

This page:
Student Work: Collages that use the subdivision of a $\sqrt{3}$ Rectangle.

Top: Kay Holloran. Cut magazine elements.

Bottom: Nancy DuClos. Torn magazine elements.

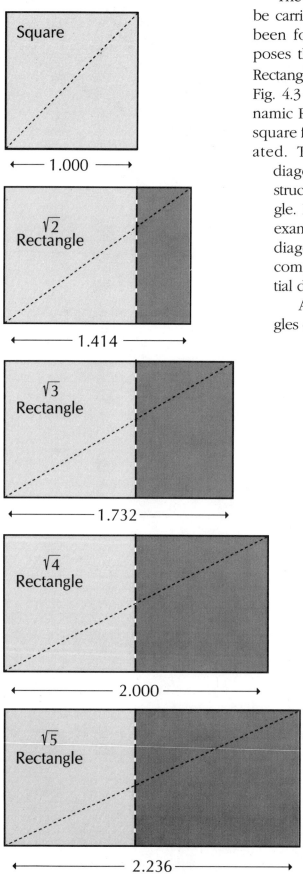

The previous construction could be carried on indefinitely, but it has been found that for practical purposes there is less use for Dynamic Rectangles beyond the $\sqrt{5}$ Rectangle. Fig. 4.3 illustrates the individual Dynamic Rectangles together with the square from which they are generated. The dotted line indicates a diagonal, necessary for the construction of the subsequent rectangle. Later in the chapter we shall examine the importance of the diagonal not only as a tool for composition but also as an essential design element.

Although the Dynamic Rectangles can always be constructed by the previous method, it may not be the most efficient one, since to construct a $\sqrt{5}$ Rectangle, you must first construct the $\sqrt{2}$, $\sqrt{3}$, and $\sqrt{4}$ respectively. Therefore, the next series of constructions provide alternate methods for obtaining the Dynamic Rectangles whereby each may be constructed separately. You are encouraged to note the relationship between the given geometric figure and the resulting Dynamic Rectangle.

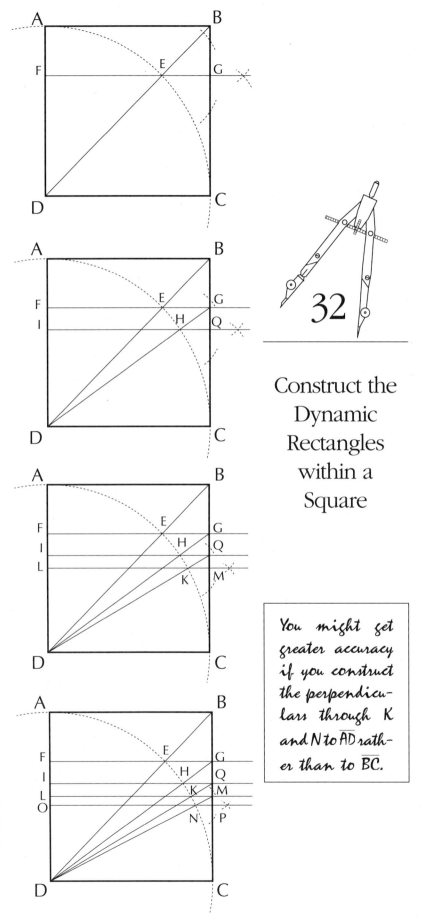

Given square ABCD

1. Place the metal tip on D and the pencil tip on A and cut arc AC.

2. Draw $\overline{DB}$, and label E the intersection of arc AC and $\overline{DB}$.

3. From E construct a perpendicular to $\overline{BC}$ and extend it so that it intersects $\overline{AD}$. Label the points of intersection F and G.

Now FDCG is a $\sqrt{2}$ Rectangle.

4. Draw $\overline{DG}$, and label H the intersection of arc AC and $\overline{DG}$.

5. From H construct a perpendicular to $\overline{BC}$ and extend it so that it intersects $\overline{AD}$. Label the points of intersection I and Q.

Now IDCQ is a $\sqrt{3}$ Rectangle.

6. Draw $\overline{DQ}$, and label K the intersection of arc AC and $\overline{DQ}$.

7. From K construct a perpendicular to $\overline{BC}$ and extend it so that it intersects $\overline{AD}$. Label the points of intersection L and M.

Now LDCM is a $\sqrt{4}$ Rectangle.

8. Draw $\overline{DM}$, and label N the intersection of arc AC and $\overline{DM}$.

9. From N construct a perpendicular to $\overline{BC}$ and extend it so that it intersects $\overline{AD}$. Label the points of intersection O and P.

Now ODCP is a $\sqrt{5}$ Rectangle.

32

Construct the Dynamic Rectangles within a Square

You might get greater accuracy if you construct the perpendiculars through K and N to $\overline{AD}$ rather than to $\overline{BC}$.

Harmonic Subdivisions of the Square Using Ø

The square may also be subdivided harmoniously into Ø proportions by the use of the Golden Cut and various diagonals. Several examples follow with information concerning their construction. You will find that points of intersection of diagonals and segments parallel to the sides play an important role in these designs. You may assume in each case that the segments that appear to be either parallel or perpendicular to each other, in fact, are.

Top left:
Fig. 4.4 C and E are the Golden Cuts of $\overline{AB}$ and $\overline{BD}$, respectively. Diagonals $\overline{BF}$ and $\overline{EF}$ are drawn. Square BCGE is drawn, and the process continues with the next square being determined by the intersection, point I, of $\overline{CH}$ and $\overline{EF}$.

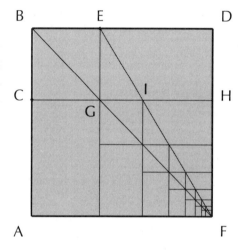

Top right:
Fig. 4.5 C and D are the Golden Cuts of $\overline{AB}$. (Remember, you can make the cut from either endpoint!) E and F are the Golden Cuts of $\overline{AC}$ and $\overline{DB}$, respectively.

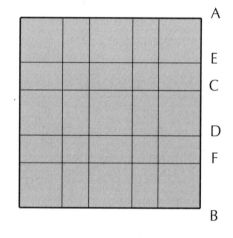

Bottom left:
Fig. 4.6 B and D are the Golden Cuts of $\overline{AC}$. F is the midpoint of $\overline{EC}$. G and H are the Golden Cuts of $\overline{EF}$ and $\overline{FC}$, respectively.

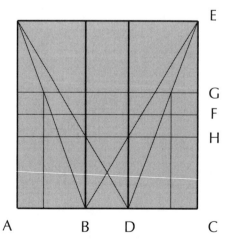

Bottom right:
Fig. 4.7 B is the Golden Cut of $\overline{AD}$. E and F are the Golden Cuts of $\overline{CD}$.

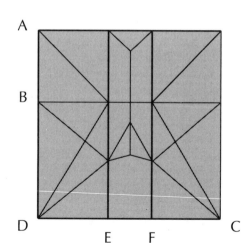

This page:
Donna Fowler. Student Work. Acrylic paint on canvas. 24" x 24".
Visual pun on a painting, Composition with Red, Yellow, and Blue, *done in 1925 by Dutch artist Piet Mondrian. This student work uses an harmonic armature. Done with preparatory drawings reconstructing the spatial divisions of the original artwork, representational images were integrated into the abstraction. The assignment required that the painting include a checkerboard floor, gray walls, the three primary colors of yellow, red, and blue, plus fabric pieces.*

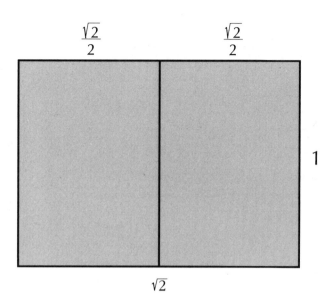

$$\frac{\sqrt{2}}{2} \qquad \frac{\sqrt{2}}{2}$$

1

$\sqrt{2}$

This page:
Top:
Fig. 4.8 Book format.

Fig. 4.9 Some natural objects may be enclosed neatly by a √2 Rectangle, such as those found in the illustrations.

Center left:
Peanut, two √2 Rectangles

Center right:
Filbert, front view

Bottom left:
Yellow Water Lily

Bottom right:
Oak Leaf

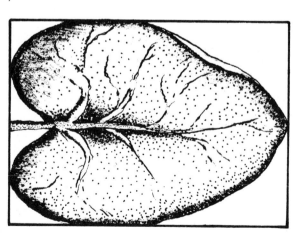

The $\sqrt{2}$ Rectangle

The simplest way to construct the √2 Rectangle is to begin with the square as in Construction 31. Therefore, no other construction is needed or offered.

The $\sqrt{2}$ Rectangle has a unique property that makes it extremely practical for the graphic artist. It is the *only* rectangle in which one half of the figure is similar to the whole. To the book designer this rectangle is ideal. For if the sides of a book are in the ratio √2:1 this means that when the book is open it has the same shape as when it is closed. In Fig. 4.8, $\frac{\sqrt{2}}{1}$ is the ratio of the length to the width of the open book. $\frac{1}{\sqrt{2}/2}$ is the ratio of the length to the width of a page (or the closed book). The following equations show these to be the same ratio.

$$\frac{1}{\sqrt{2}/2} = \frac{2}{\sqrt{2}}\frac{\sqrt{2}}{\sqrt{2}} = \frac{2\sqrt{2}}{2} = \sqrt{2}.$$

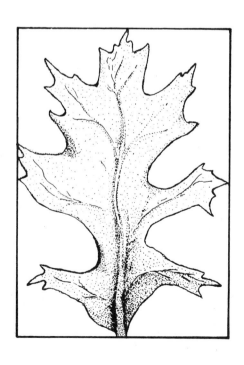

According to Paul Grillo, in his book *Form, Function and Design,* the $\sqrt{2}$ Rectangle has become the standard for most of the windows of the villages and towns of Europe as well as being the standard proportion for the window pane. It is also used in the design of office furniture and file cabinets as it makes it possible to divide any drawer in half without wasted space or change in proportion.

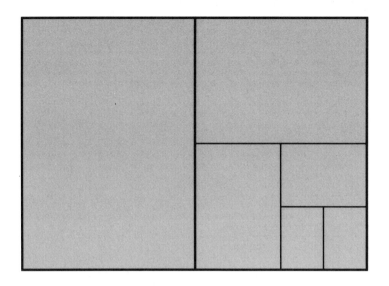

Top:
Fig. 4.10 The diagram illustrates a subdivision of the $\sqrt{2}$ Rectangle which is gotten by dividing the rectangle in half through the midpoint of the longer side and repeating the process.

Center:
Lion fish fits within a $\sqrt{2}$ Rectangle.

Bottom:
Fig. 4.11 The $\sqrt{2}$ Rectangle is found in an unexpected place—a wooden blanket box of the Northwest Indian tribe, the Tsimshians. It reiterates the tendency toward breaking a line segment into Ø proportions, as C is very close to the Golden Cut of $\overline{AB}$. This tendency may be more pronounced in such a culture so closely attuned to the natural world.

129

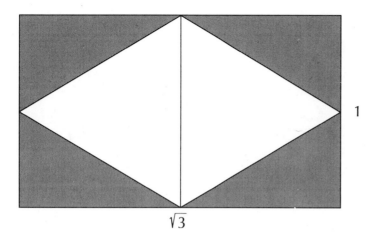

The √3 Rectangle

The √3 Rectangle may be generated in several ways, each being useful for the artist-designer. If two equilateral triangles are placed side by side and enclosed in a rectangle, it will be a √3 Rectangle. This property is the basis for the following constructions, even though the two triangles are not actually drawn.

If opposite sides of a **regular hexagon** are joined by parallel line segments from the vertices, three intersecting √3 Rectangles are formed, producing a smaller regular hexagon in the center. Fig 4.13 illustrates this process repeated three times. It may be carried out indefinitely (or at least until eyes, fingers and tools no longer function!).

If, in Construction 33, points A, F, D, B, C and E were all connected sequentially by line segments, the resulting figure would be a regular hexagon. Therefore, $\overline{FD}$ and $\overline{EC}$ would be opposite sides as described, and FDCE must be a √3 Rectangle.

Top:
Fig. 4.12 Equilateral Triangles in a √3 Rectangle.

Center:
Animal face drawing by Ann MacLean based on Fig. 4.13. Technical pen and ink on paper.

Bottom right:
Fig. 4.13   Repeated subdivisions of a regular hexagon produces smaller and smaller sets of intersecting √3 Rectangles.

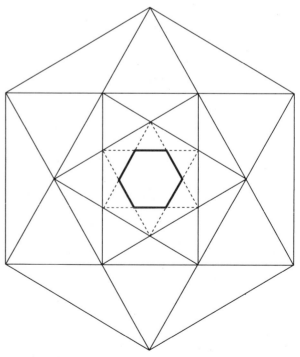

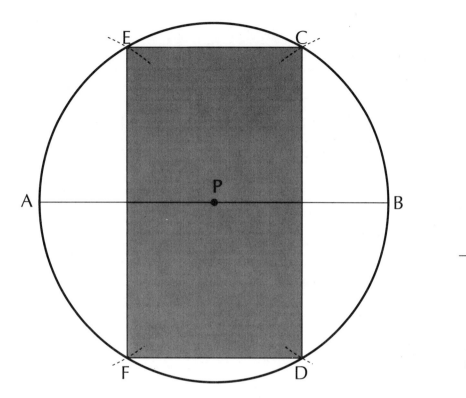

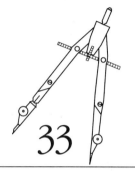

33

Construct a √3 Rectangle Given a Circle

Given Circle P

1. Draw a diameter and label its endpoints A and B, respectively.

2. With the compass open to the radius of the circle, place the metal tip on B and cut arcs on either side of $\overleftrightarrow{AB}$. Label the points of intersection C and D, respectively.

3. Without changing the setting, place the metal tip on A and cut arcs on either side of $\overleftrightarrow{AB}$. Label points of intersection E and F, respectively.

4. Draw $\overline{EF}$, $\overline{FD}$, $\overline{DC}$, and $\overline{EC}$.

Now FDCE is a √3 Rectangle.

Note: The position of the diameter is arbitrary. For the rectangle to lie horizontally the diameter should be drawn vertically. A diameter drawn diagonally will yield a rectangle that lies diagonally perpendicular to the diameter.

131

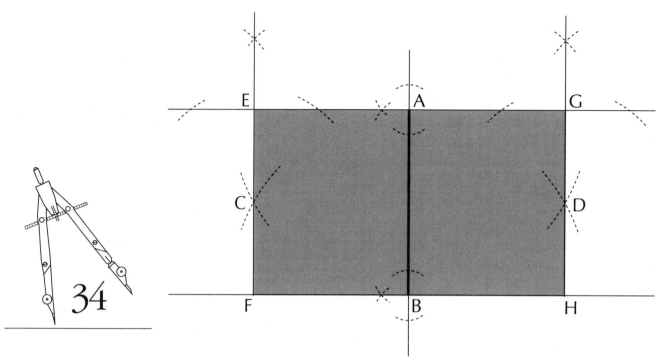

34

Construct a $\sqrt{3}$ Rectangle Given the Width

Given $\overline{AB}$

1. Extend $\overleftrightarrow{AB}$ in both directions. Construct perpendiculars to $\overleftrightarrow{AB}$ through A and B, respectively.

2. Place the metal tip on A and the pencil tip on B and cut arcs on either side of $\overleftrightarrow{AB}$. Without changing the setting, place the metal tip on B and cut arcs that intersect the ones just drawn. Label the points of intersection C and D, respectively.

3. From C construct a perpendicular to the line through A and extend it so that it intersects the line through B. Label the points of intersection E and F, respectively.

4. Repeat Step 3 at D. Label the points of intersection G and H, respectively.

Now EFHG is a $\sqrt{3}$ Rectangle.

The √3̄ Rectangle is less often connected with organic forms but
abounds in crystalline ones, like
snowflakes, since it is inextricably bound to the hexagon.

In two-dimensional de-
sign, some consider it discor-
dant to mix the themes of √2̄
and √3̄ together on the same
surface. However, each may be
used with the square to create
harmonious compositions. In three dimensions, though, the artist might use
different root rectangle themes on
different faces, or planes, provided
that the design theme is consistent
within each plane. Fig. 4.14, below,
shows one example in which Nature
uses this principle of design.

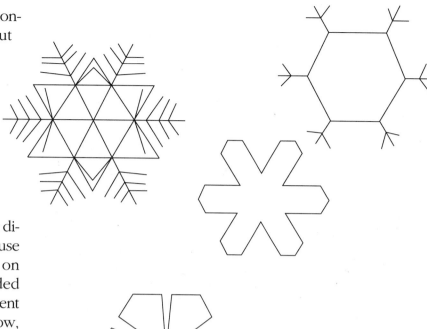

Note: It is interesting that the √3̄ Rectangle is the one that Plato found to be the most beautifully proportioned. However, since it is rarely found in the artwork of that period, it seems his peers did not agree with his choice.

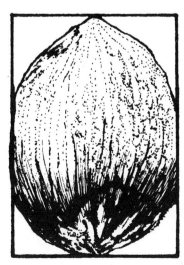

√2̄ Rectangle

√3̄ Rectangle

Left:
Fig. 4.14 Front and side views of a filbert nut. This natural form fits within implied rectangular frames.

The Phi-Family Rectangles

The Golden Rectangle, with which you are now quite familiar, gives us a family of rectangles that works together with great compatibility for use in plane design. If the width of each of these rectangles measures 1, then the lengths measure 1 (square); $\sqrt{\varnothing}$ (1.272); $\varnothing$ (1.618); $\sqrt{4}$ (2); $\sqrt{5}$ (2.236); and $\varnothing{+}1$ (2.618), respectively, where the numerical values in parentheses are rounded to the nearest thousandth. We will be discussing each of these rectangles individually later on in this chapter, but we have provided the diagram of the family tree to point out the relationship among the members of the family.

Above:
Linda Maddox. Composition using members of the Ø-Family Rectangles. Technical pen and ink on paper.

Below:
Fig. 4.15 Ø-Family Rectangles

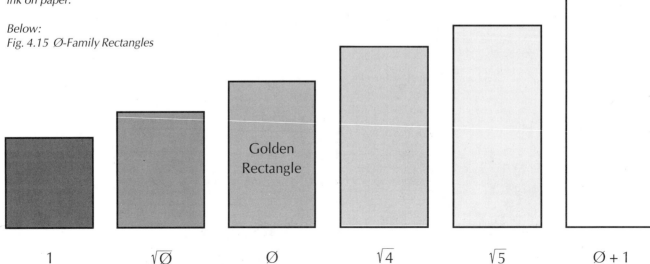

| 1 | $\sqrt{\varnothing}$ | $\varnothing$ | $\sqrt{4}$ | $\sqrt{5}$ | $\varnothing + 1$ |

This page:
Fig. 4.16   PHI-FAMILY TREE
Two members of this family are
root rectangles. They are the
√4 Rectangle and the √5 Rec-
tangle. It should be noted that
the √4 Rectangle, being a dou-
ble square, would also be
compatible with either the √2
Rectangle or the √3 Rectangle
in visual composition.

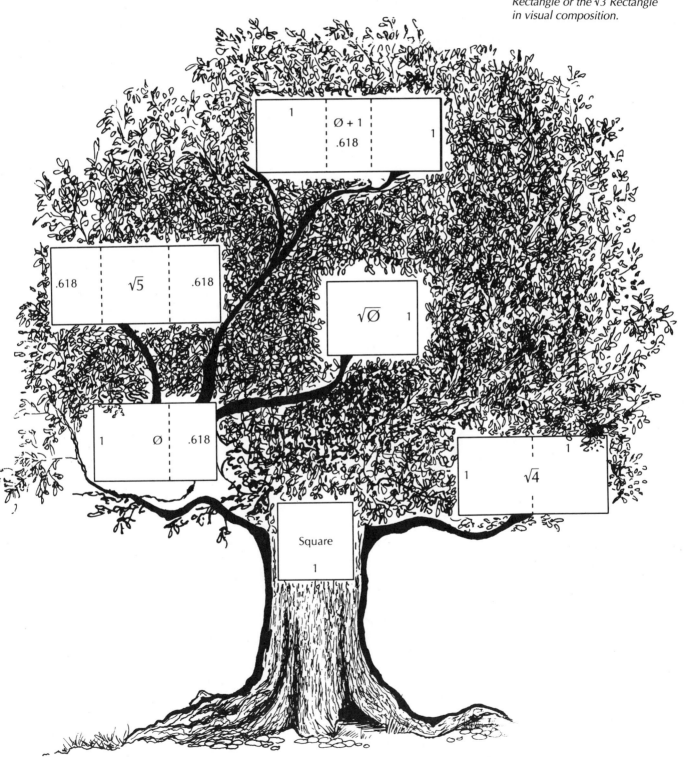

Ø-Family: √Ø Rectangle

The second rectangle in progression, according to the length of the sides, is the √Ø Rectangle; the first being the unit square discussed previously. The √Ø Rectangle is sometimes called the Rectangle of Price since it is derived from the Triangle of Price. Fig. 4.17 illustrates the relationship. You will recall that a Triangle of Price can be constructed from a Golden Rectangle (see Construction 21, Chapter 3). Therefore, it is logical to construct a √Ø Rectangle from a Golden Rectangle as in Construction 35.

Most of us handle examples of √Ø Rectangles each day without being aware of the golden connections. The standard size of notebook and typing paper (8 1/2" x 11") very closely approximates the dimensions of the √Ø Rectangle. Other larger standard sizes of paper also have dimensions in the same ratio, as illustrated in Fig. 4.18.

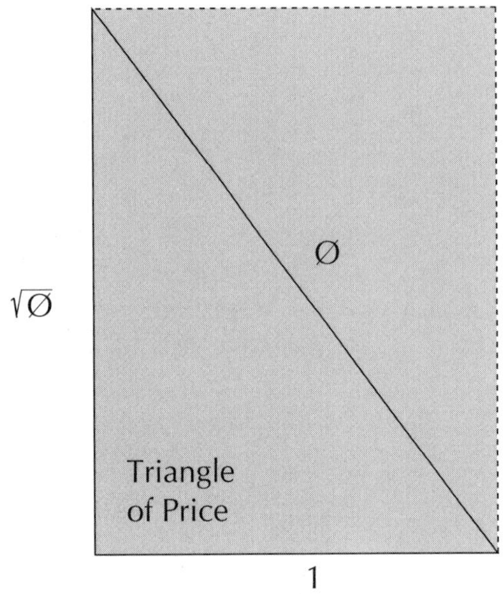

√Ø

Ø

Triangle
of Price

1

Top:
Fig. 4.17 √Ø Rectangle.

Bottom:
Fig. 4.18 Commercial paper sizes.

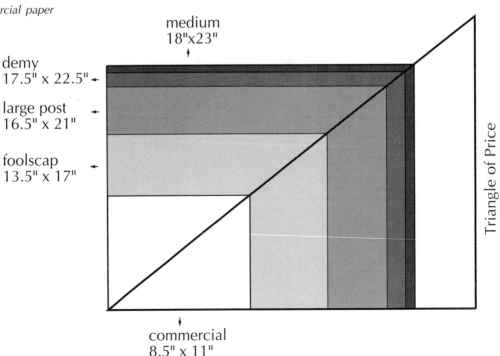

medium
18"x23"

demy
17.5" x 22.5"

large post
16.5" x 21"

foolscap
13.5" x 17"

commercial
8.5" x 11"

Triangle of Price

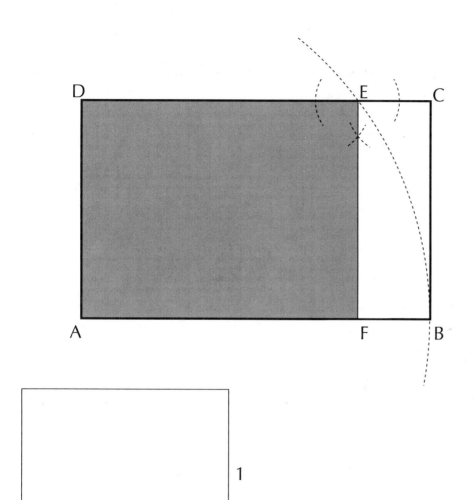

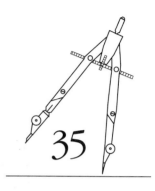

35

Construct a
√Ø Rectangle
Given a
Golden
Rectangle

Given Golden Rectangle ABCD

1. Place the metal tip on A and the
pencil tip on B and cut an arc that
intersects $\overline{DC}$. Label the point of intersection E.

2. Construct a perpendicular from E
to $\overline{AB}$. Label the point of intersection
F.

Now AFED is a √Ø Rectangle.

137

Ø-Family: √4 Rectangle

Although the Golden Rectangle is the next in progression, it has already been discussed in Chapter 2. Therefore, we will go on to the √4 Rectangle. Since √4 = 2, any rectangle that is twice as long as it is wide is a √4 Rectangle. The fact that the length is double the width makes this rectangle a double square, and constructing it is an easy affair. See Construction 36. The √4 Rectangle is the Japanese modular unit, the tatami, from which all Japanese architectural proportion is derived. A room is built around a rectangle which is formed by the arrangement of a particular number of tatami mats. The shape and size of the house itself depends upon the distribution of the room modules.

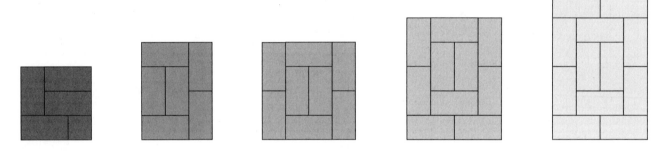

Top:
Fig. 4.19 Here are some possible arrangements of tatami mats.

Bottom:
Fig. 4.20 Japanese house plan.

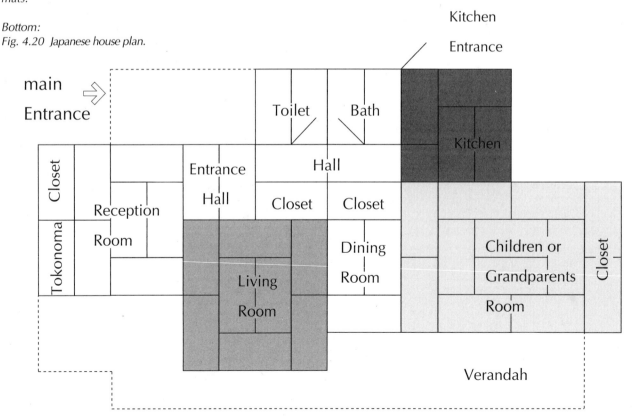

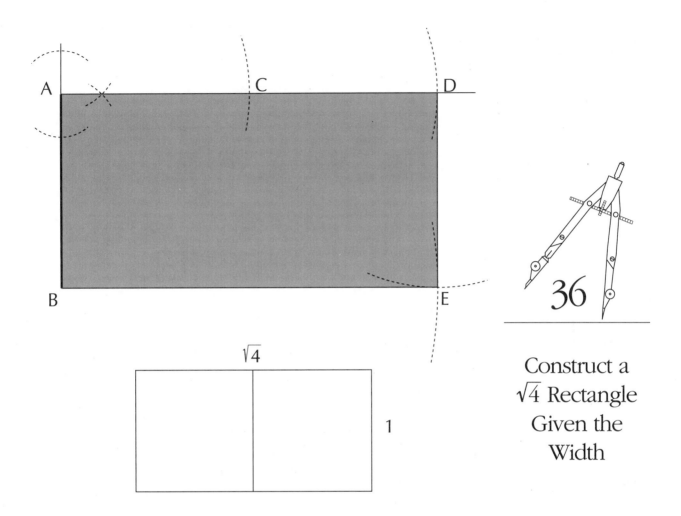

√4

1

36

Construct a
√4 Rectangle
Given the
Width

Given $\overline{AB}$

1. Extend $\overrightarrow{BA}$, and construct a perpendicular to $\overleftrightarrow{AB}$ through A.

2. Place the metal tip on A and the pencil tip on B and cut an arc that intersects the line drawn in Step 1. Label the point of intersection C.

3. Without changing the setting, place the metal tip on C and cut a second arc on $\overrightarrow{AC}$. Label the point of intersection D.

4. Without changing the setting, place the metal tip on D and cut an arc in the interior of ∠DAB.

5. Open the compass so that it measures AD. Place the metal tip on B and cut an arc that intersects the one drawn in Step 4. Label the point of intersection E.

6. Draw $\overline{BE}$ and $\overline{DE}$.

Now ABED is a √4 Rectangle.

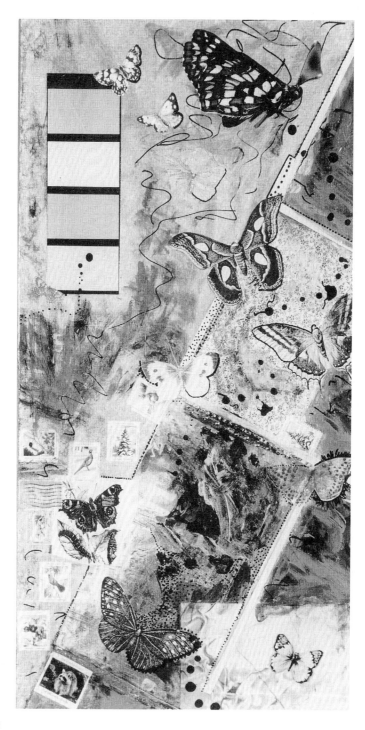

This page:
Rochelle Newman. Flights of
Fancy. *Mixed media, paint,*
ink, and found images in a $\sqrt{4}$
Rectangle format. One unit of
a collaborative work by the
members of SLMM, the Soci-
ety of Layerists in Multi Media.

Ø-Family: $\sqrt{5}$ Rectangle

Next to the Golden Rectangle, the $\sqrt{5}$ Rectangle is perhaps the most widely used both by humans and Nature as a structure for design. The $\sqrt{5}$ Rectangle is derived by overlapping two Golden Rectangles so that their common area is a square. This property allows for easy construction of a $\sqrt{5}$ Rectangle simply by extending the construction for a Golden Rectangle.

Looking once again at the Parthenon (Fig. 4.21), we can see the use of the √5 Rectangle together with the square, in the floor plan and within the enclosing Golden Rectangle of the facade.

In a very different art form, by a very different cultural group, we find the √5 Rectangle once again (Fig. 4.22). Whether a conscious choice or an intuitive one, the √5 Rectangle is used extensively in the blanket designs of the Northwest Coast Indian tribes with aesthetically satisfying results.

Below:
Fig. 4.21 The facade and floor plan of the Parthenon.

Lower left:
Fig 4.22 Drawings after a Northwest Coast Tlingit Chilkat Indian Blanket.

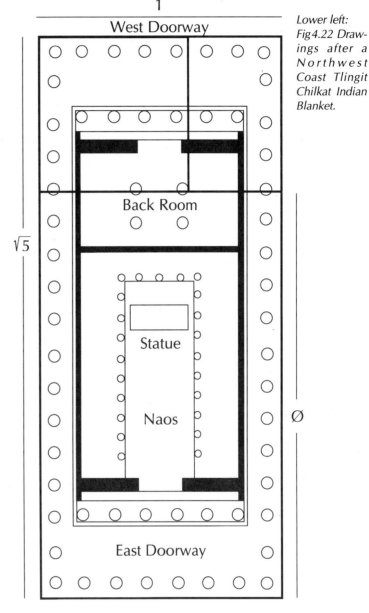

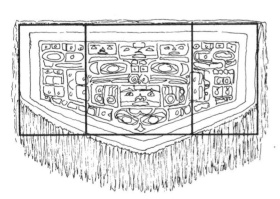

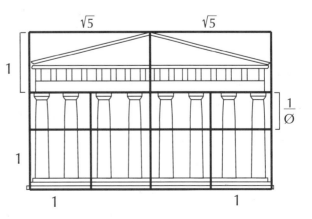

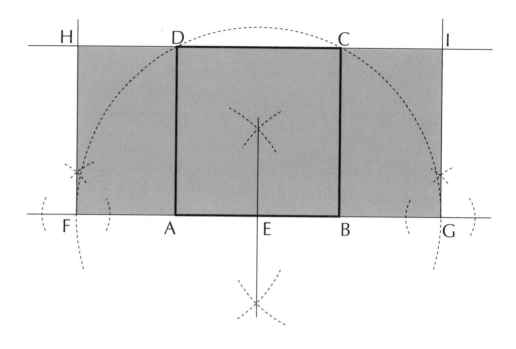

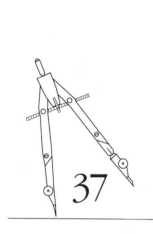

37

Construct a √5 Rectangle Given a Square

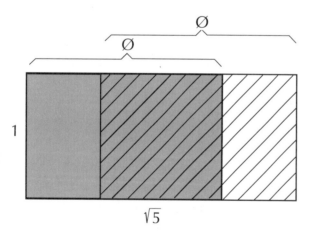

Two overlapping Golden Rectangles form the √5 Rectangle.

Given square ABCD

1. Extend $\overleftrightarrow{AB}$ and $\overleftrightarrow{CD}$. Bisect $\overline{AB}$ and label the midpoint E.

2. Place the metal tip on E and the pencil tip on C and cut an arc that intersects $\overleftrightarrow{AB}$ twice. Label the points of intersection F and G, respectively.

3. Construct perpendiculars to $\overleftrightarrow{AB}$ at F and G and extend them so that they intersect $\overleftrightarrow{DC}$. Label the points of intersection H and I, respectively.

Now FGIH is a √5 Rectangle.

Note: AGID and FBCH are both Golden Rectangles.

Ø-Family: Ø+1 Rectangle

The other rectangle in the Ø-Family is the one we have christened the Ø+1. It is obtained by adding a square to one end of an existing Golden Rectangle. Like the $\sqrt{5}$, the Ø+1 Rectangle contains two overlapping Golden Rectangles. This time they share the small Golden Rectangle rather than the square. Therefore, the construction of the Ø+1 Rectangle is an easy extension of the construction of the Golden Rectangle.

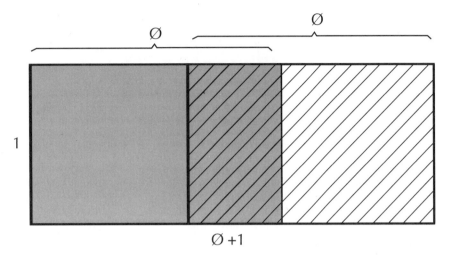

Top:
Fig. 4.23 Two overlapping Golden Rectangles form the Ø+1 Rectangle.

Bottom: Rochelle Newman. Two views of one of a series of Ø-Family rectangular solids. The sides of this one are $\sqrt{5}$ Rectangles and squares. Acrylic paint on watercolor paper.

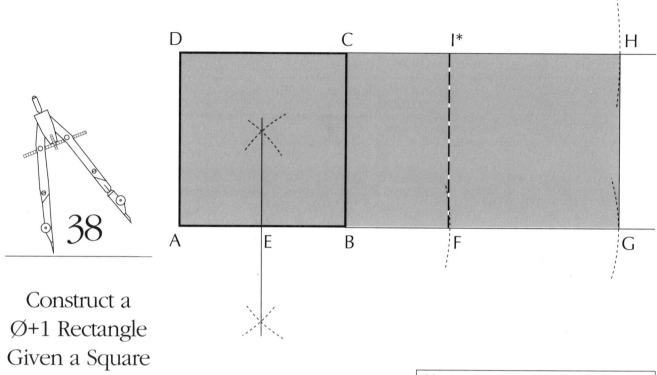

38

Construct a Ø+1 Rectangle Given a Square

Ø + 1

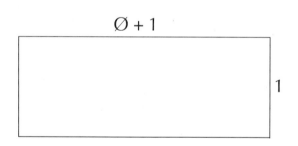

1

Given square ABCD

1. Extend $\overrightarrow{AB}$ and $\overrightarrow{DC}$. Bisect $\overline{AB}$ and label the midpoint E.

2. Place the metal tip on E and the pencil tip on C and cut an arc that intersects $\overrightarrow{AB}$. Label the point of intersection F.

3. Open the compass to measure AB. Then, with the metal tip on F, cut an arc on $\overrightarrow{AF}$ on the side of F opposite A. Label the point of intersection G.

4. Open the compass to measure AG. Then, with the metal tip on D, cut an arc on $\overrightarrow{DC}$. Label the point of intersection H.

5. Draw $\overline{GH}$.

Now AGHD is a Ø+1 Rectangle.

Note: If a perpendicular is drawn from F to I on $\overline{DH}$, AFID and BGHC are both Golden Rectangles.

This page:
Fig. 4.24 Implied rectangular
frames suggest limits on the
growth of natural forms.

Natural Forms Escribed by Ø-Family Rectangles

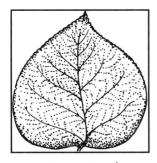

Aspen Leaf
Square

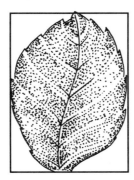

Strawberry Leaf
√Ø Rectangle

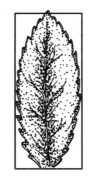

Raspberry Leaf
√5 Rectangle

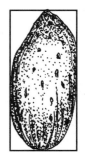

Almond
√4 Rectangle

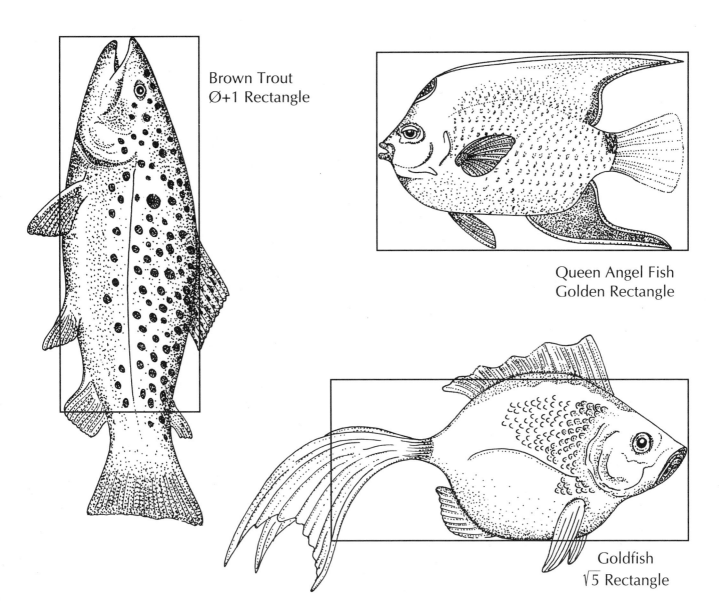

Brown Trout
Ø+1 Rectangle

Queen Angel Fish
Golden Rectangle

Goldfish
√5 Rectangle

145

Top:
Fig. 4.25 Illustration of the Ø connections in several golden figures:
A. two intersecting pentagons give the illusion of a decagon,
B. Golden Triangles,
C. Lutes of Pythagoras with their corresponding pentagrams, and
D. the √5 Rectangle composed of two overlapping Golden Rectangles.

Bottom left and right:
Linda Maddox. Variations on the diagram in Fig. 4.25. Technical pen and ink on paper.

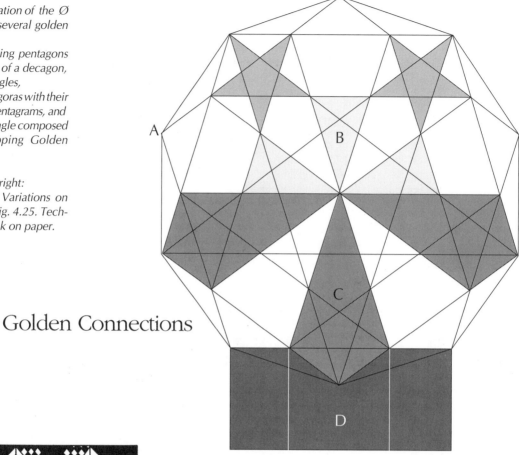

Golden Connections

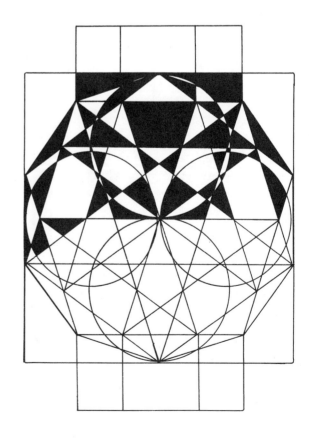

The Dynamic Parallelograms

With a little mathematical sleight-of-hand we can transform the Dynamic Rectangles into geometric figures that have even more qualities of implied motion. We have named these figures the **Dynamic Parallelograms**.

A parallelogram is formed whenever opposite sides of a quadrilateral are parallel. There are no restrictions on the angles as with a rectangle. However, it turns out that consecutive angles are always **supplementary**. This means that a rectangle can be "tipped" such that the sides remain the same length but one angle of a consecutive pair will meas-ure between 0° and 90° and the other will measure between 90° and 180°, and their sum will be 180°.

Fig. 4.26 illustrates a few of the infinite number of √3 Parallelograms that can be constructed from a √3 Rectangle. Using the width of the rectangle as the radius, arcs, centered at each endpoint of a side (D and C), are drawn. By connecting one vertex to any point on one of the arcs (say A), the side of the parallelogram retains the length of the side of the original rectangle. A segment parallel to the opposite side is then drawn and extended until it intersects the other arc ($\overline{AB}$). The final side is then drawn by connecting the point of intersection to the original vertex ($\overline{BC}$). The same procedure could be followed for other Dynamic Rectangles. Note that the Dynamic Rectangles are themselves members of the family of Dynamic Parallelograms.

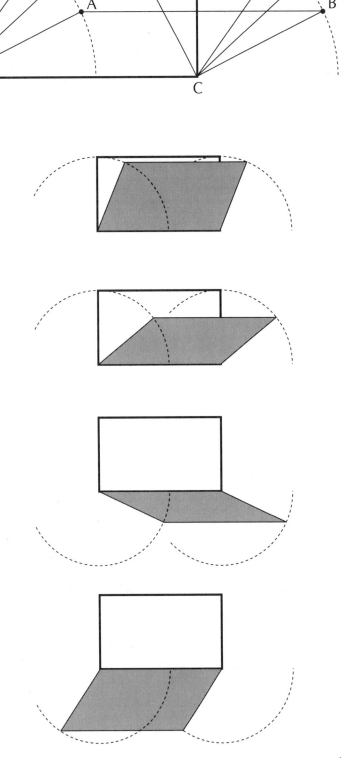

147

The Reciprocal

Each of the rectangles we have just discussed provides a form for designing harmoniously. The designer, however, needs more than just the outer ratios. He or she needs to be able to subdivide the rectangle into smaller units that continue to preserve the overall harmony. One method of subdividing a rectangle is by the use of the reciprocal, which is a similar rectangle cut from the parent, whose length is the width of the parent rectangle.

In Fig. 4.27, the shaded rectangle is the reciprocal of ABCD. Since the two rectangles are similar, they have the same shape (the position, of course, is altered) and their sides are in proportion. That is, $\frac{AB}{BC} = \frac{BC}{EC}$. The construction for the reciprocal of a rectangle follows on the next page.

Right:
Fig. 4.27 The reciprocal of a rectangle. ABCD ~ BCEF.

Below:
Fig. 4.28 Construction 39 can be used to find the reciprocals of any rectangle. It is only in the family of Dynamic Rectangles that multiples of the reciprocal fit neatly into the parent with no leftover spaces. You will notice that as the length of the parent rectangle increases, the relative size of the reciprocal decreases. The √2 Rectangle is the only one in which one half of the figure is similar to the whole. In each case ABCD ~ FEBC.

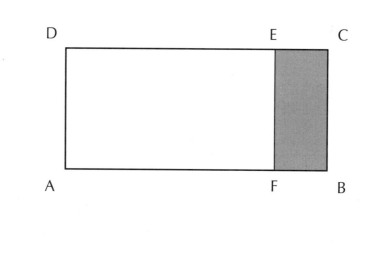

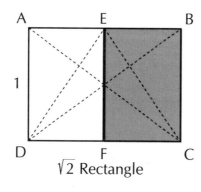

√2 Rectangle

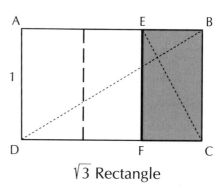

√3 Rectangle

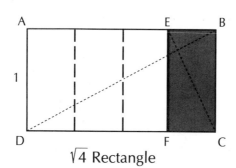

√4 Rectangle

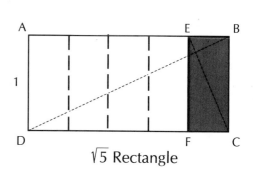

√5 Rectangle

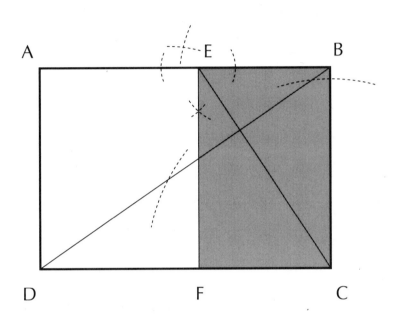

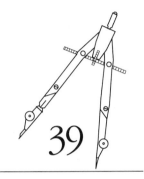

39

Construct the
Reciprocal of a
Rectangle

Given rectangle ABCD

---

1. Draw $\overline{DB}$.

2. From C construct a perpendicular to $\overleftrightarrow{DB}$ and extend it to intersect $\overline{AB}$. Label the point of intersection E.

3. From E construct a perpendicular to $\overleftrightarrow{DC}$. Label the point of intersection F.

Now FEBC is the reciprocal of ABCD, and ABCD~FEBC.

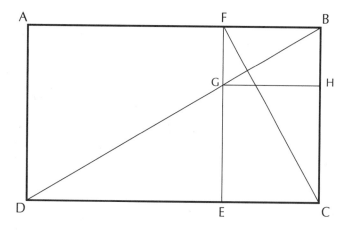

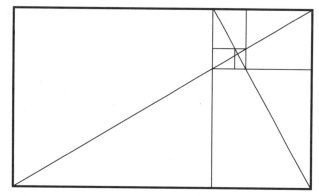

You will notice the importance of the diagonals of both the parent and reciprocal rectangles in Construction 39. These two intersect each other at right angles and provide the framework to construct smaller and smaller similar rectangles. It is this property of similarity that relates the reciprocal to the phenomenon of growth in Nature. Natural growth patterns also have to do with similarity and conservation of space.

In all the Dynamic Rectangles, the length of the parent rectangle is a multiple of the width of the reciprocal. This means that the rectangle can be subdivided into similar shapes with no leftover area (Fig. 4.28). This fact enables you to design with both economy and unity without relying on simple repetition.

Top:
Fig. 4.29 By Construction 39 we know that FECB is the reciprocal of ABCD. Then by drawing in $\overline{GH}$ at the point of intersection of $\overline{DB}$ and $\overline{EF}$, we have that GHBF is the reciprocal of EFBC. So ABCD ~ EFBC ~ FBHG.

Center:
Fig. 4.30 The diagonals used to construct the reciprocal also provide the framework for constructing subsequent reciprocals as the rectangles get smaller and smaller.

Bottom:
Fig 4.31 The fact that in the Dynamic Rectangles the reciprocal can be fitted neatly into the parent rectangle allows for many varied design possibilities. For demonstration purposes we have used only the √3 Rectangle and its reciprocal, together with various diagonals. However, the same principles of division would hold for the other Dynamic Rectangles as well.

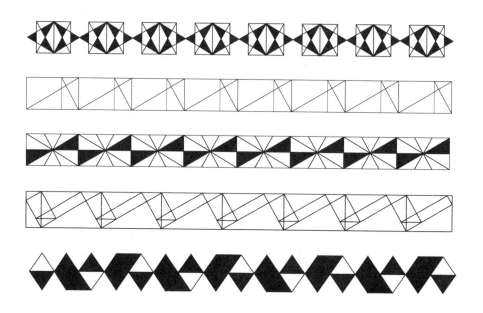

The Gnomon

Within any given rectangle, there exists the reciprocal together with its complement, the **gnomon** (pronounced no-men). This term was first used by the Babylonians to name an upright stick placed in the ground. It functioned as a crude sundial. For the Greeks, the term came to represent a carpenter's square, a shape similar to that in the first diagram in Fig. 4.32. The concept was further expanded and is now taken to mean *any shape that can be added to a given shape to produce a similar shape.* In Fig. 4.32, the gnomon is the shaded portion. When this is added to the unshaded portion, a figure similar to the unshaded portion is obtained.

As you can see from these examples, a single figure may have gnomons of many different sizes and shapes, and they need not reflect the shape of the original figure. We see a rectangle whose gnomon is in the shape of a carpenter's square and one whose gnomon is another rectangle. One triangle has a V-shaped gnomon while the other one is in the shape of a trapezoid.

*This page:
Fig. 4.32 Gnomons. In each case, the shaded portion is the gnomon of the unshaded portion. The overall figure is similar to the unshaded portion.*

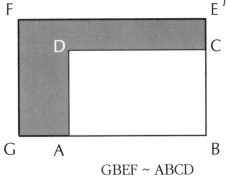

GBEF ~ ABCD

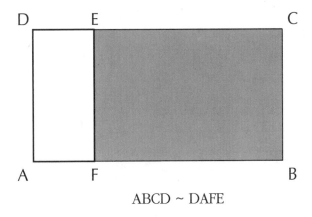

ABCD ~ DAFE

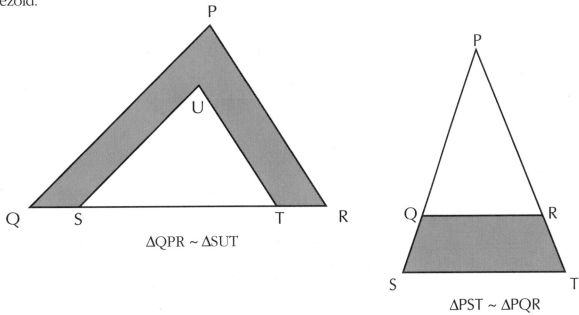

ΔQPR ~ ΔSUT

ΔPST ~ ΔPQR

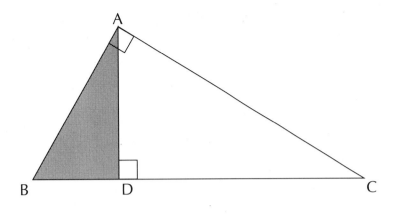

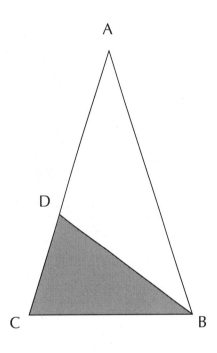

Top:
Fig. 4.33 In the diagram we see that △ADC ~ △BDA ~ △BAC. Since all three triangles are similar, either of the two smaller triangles when added to the other produces a similar triangle.

Center:
Fig. 4.34 △ABC ~ △BCD and △ADB is the gnomon of △BDC.

Bottom:
Fig. 4.35 A series of gnomons in the Golden Triangle.

A triangle may also have a gnomon that is triangular in shape. In fact, an interesting case occurs in right triangles. If an altitude is dropped from the right angle, two triangles are formed, each of which is the gnomon of the other (Fig. 4.33). Any triangle can be subdivided so that one part is the gnomon of the other. (You will have the opportunity to do this in the problem set at the end of the chapter. Lucky you!)

The Golden Triangle, however, is a unique and elegant case. You will recall from the previous chapter that the Golden Triangle is the isosceles triangle whose angles measure 36°, 72°, 72° respectively. Since the vertex angle measures exactly one half a base angle, we can bisect one of the base angles to form another Golden Triangle within the original. (See Figs. 4.34 and 4.35.) This process can be repeated to produce a series of smaller and smaller Golden Triangles and the resulting figure forms another structure for a logarithmic-like spiral which we shall examine in Chapter 6.

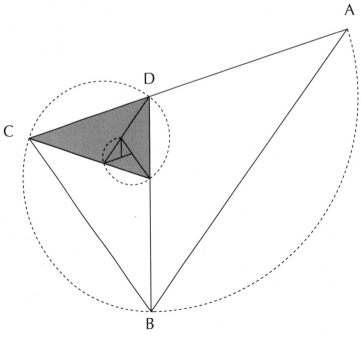

The process of adding a gnomon to a form can be carried on indefinitely giving us Nature's law of continued similarity. This is especially exemplified in growth patterns of structures such as shells and horns which, again, we shall study further in Chapter 6. We recognize many trees by their overall shapes, each species having characteristic growth patterns. It seems then that, allowing for the deviations caused by natural forces, the new growth each season is the gnomon for the tree's shape of the previous year (Fig. 4.36).

The concept of gnomon can also be applied to numbers. Although we tend to think of numbers as abstractions to name quantities, it is possible to examine them within a geometric framework. In fact, the Pythagoreans were intrigued by classifying numbers according to the shape that would most economically hold the given number of units. The Greeks could more easily identify with the idea of the number of units in a quantity, since they indicated a number by the use of pebbles or marks in the sand.

This page:
Fig. 4.36 Natural gnomons.

Top to Bottom:
Chambered Nautilus
Northern White Cedar
Stump of a Baldwin Apple Tree

153

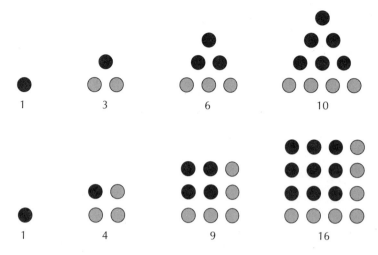

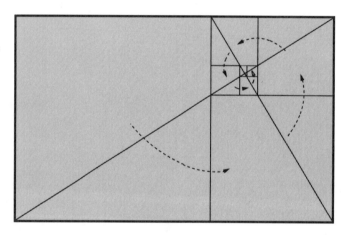

Top:
Fig. 4.37 The first four triangular numbers and square numbers. Both the numbers 4 and 10 were worshipped as divinities by the Pythagoreans.

Below:
Fig. 4.38 The Whirling Square Rectangle.

In Fig. 4.37, the row lightly shaded becomes the gnomon for the previous triangular number. If we look at the numbers named by these rows, we see that they are 2, 3, and 4. If the process is carried on we have gnomons named by 5, 6, 7, 8, etc. Thus, the set of gnomons for the triangular numbers is the set of natural numbers with the exception of 1.

The numbers we know as perfect squares are so named because the Pythagoreans saw them in the shape of squares. Again, the gnomon for the preceding square number is seen in the lightly shaded area in the shape of the carpenter's square. This time we have the sequence 3, 5, 7, etc., or the set of odd numbers with the exception of 1.

Let us return to the concept of gnomons in rectangles and look at the now familiar Golden Rectangle which is a special case. Chapter 2 demonstrated that any number of Golden Rectangles can be constructed simply by adding a square to the length. This means that the square is the gnomon of the Golden Rectangle and it is the *only* rectangle with that unique property. In fact, the Golden Rectangle is sometimes referred to as the **Whirling Square Rectangle,** since upon repeated division of the rectangle into reciprocal and square gnomon, the squares appear to whirl around the point of intersection of the main diagonal and the diagonal of the reciprocal (Fig. 4.38).

The Armature of the Rectangle

To this point we have dealt primarily with non-objective design. However, throughout the history of art, many painters were, and are, concerned with representing three-dimensionality on a two-dimensional surface. For this problem the artist needs a framework to provide the scaffolding on which to erect his or her composition, be it abstract, representational, planar, or three-dimensional. When working with complex content and form, more than random placement is necessary for strength of design.

Twentieth century painter Charles Bouleau, in his book *The Painter's Secret Geometry,* examined art throughout the ages looking for the underlying structures of the various works. He analyzed the paintings of such masters as Vermeer, Botticelli, Cezanne, and Durer. In each case he discovered the composition related to specific rectangular divisions which are a part of what he chose to call "the armature of the rectangle."

We shall briefly touch on two important ideas concerning the division of the rectangle for the placement of key forms within a representational work. Again, the diagonal plays the most important role in the evolution of the armature. By using various diagonals, the artist-designer can subdivide the entire rectangle into a series of smaller sections that relate to the entire composition. The viewer may not be consciously aware of these diagonals but they help to create the underlying order of a work. In Fig.4.39 the evolution of an armature is shown in sequence.

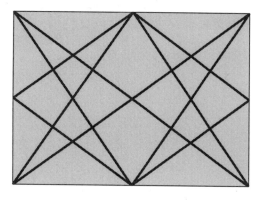

This page:
Fig. 4.39 An armature of a rectangle.

Top to bottom:
The two main diagonals are drawn.

Line segments are drawn from the midpoints of the longer sides of the rectangle to the vertices of the rectangle.

Line segments are drawn connecting consecutive midpoints of all the sides.

Line segments are drawn from the midpoints of the shorter sides of the rectangle to the vertices of the rectangle.

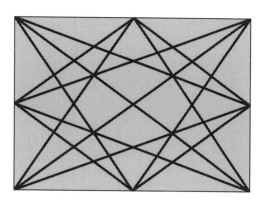

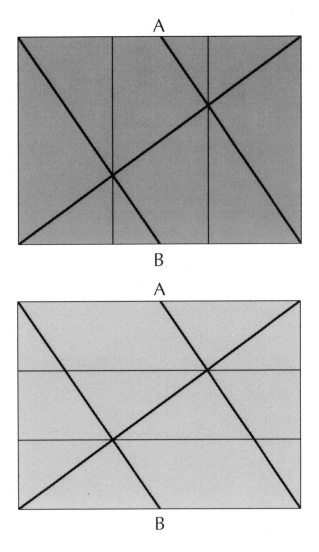

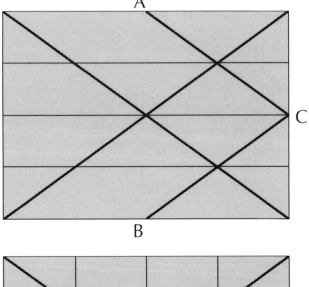

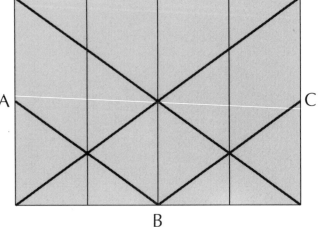

The rectangle may be divided into *thirds* by drawing parallels to the sides through the points of intersection of a main diagonal and the segments joining a vertex to the midpoints of the longer side. It may be divided into *fourths* by using points of intersection of the main diagonals and line segments joining consecutive midpoints of the sides (Fig. 4.40). Some of the points of intersection of these diagonals give us reference points for interesting subdivisions of the area which may be particularly useful when working with still life or landscape. (See analysis of painting on pages 158 and 159).

This Page:
Fig. 4.40 Armatures of rectangles using parallel subdivisions in thirds and fourths. In each case A, B and C are midpoints of the sides of the rectangles.

Top to bottom:
A vertical subdivision into thirds when parallels are drawn to the longer sides.

A horizontal subdivision into thirds when parallels are drawn to the shorter sides.

A horizontal subdivision into fourths when parallels are drawn to the shorter sides.

A vertical subdivision into fourths when parallels are drawn to the longer sides.

A way to subdivide a Dynamic Rectangle into smaller rectangles similar to itself is to divide each side into *n* equal units *(see Construction 9)* where n is any number. Connect opposite sides by line segments joining the points of division. This is illustrated in Fig. 4.41. Within these divisions diagonals can again be used to focus interest in a composition.

Although the sides of the rectangle may be divided into thirds or fourths by the previously described method, the diagonals themselves play an important role in providing the framework for a composition. The points of intersection of the various diagonals become focal points for images in the composition. Fig. 4.42 shows some of the ways the Dynamic Rectangles may be subdivided in order to provide workable armatures.

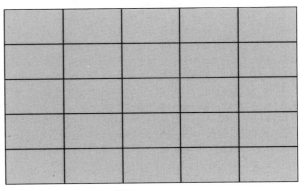

$\sqrt{3}$ Rectangle

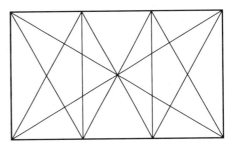

$\sqrt{3}$ Rectangle

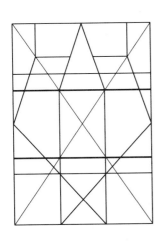

$\sqrt{2}$ Rectangle

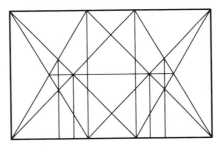

Golden Rectangle

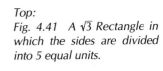

$\sqrt{5}$ Rectangle

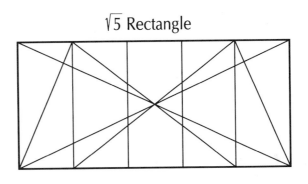

Top:
Fig. 4.41 A $\sqrt{3}$ Rectangle in which the sides are divided into 5 equal units.

Center and bottom:
Fig. 4.42 Armatures of Dynamic Rectangles.

Opposite page:
Colville, Alex. Skater. 1964. Synthetic polymer paint on composition board, 44 1/2 x 27 1/2". Collection, The Museum of Modern Art, New York. Gift of R.H. Donnelley Erdman (by exchange).

This page:
Fig. 4.43 In Skater, the outer dimensions of the painting consist of a Golden Rectangle. The horizontal and vertical lines in the armature connect Golden Cuts of the sides. Diagonals play an important role in defining shapes in space.

Although the painting, *Skater*, by the Twentieth Century Canadian artist, Alex Colville, appears to be about the spontaneous act of ice skating, it is also about the essential pictorial problem of depicting a three-dimensional experience on a bounded two-dimensional surface. In any painting, the image resides within a particular rectangle which is subdivided in a specific way. The subdivisions play a crucial part in the essence of a strong composition. What the viewer sees is only an illusion of depth created by the skillful handling of shapes. Fig. 4.43 illustrates the use of the armature in *Skater*.

This page:
Ann MacLean. Student Work.
Collage using the armature of
Skater. Found photographic
elements and hand drawing.

Dynamic Rectangles in Relation to Special Triangles and Other Plane Figures

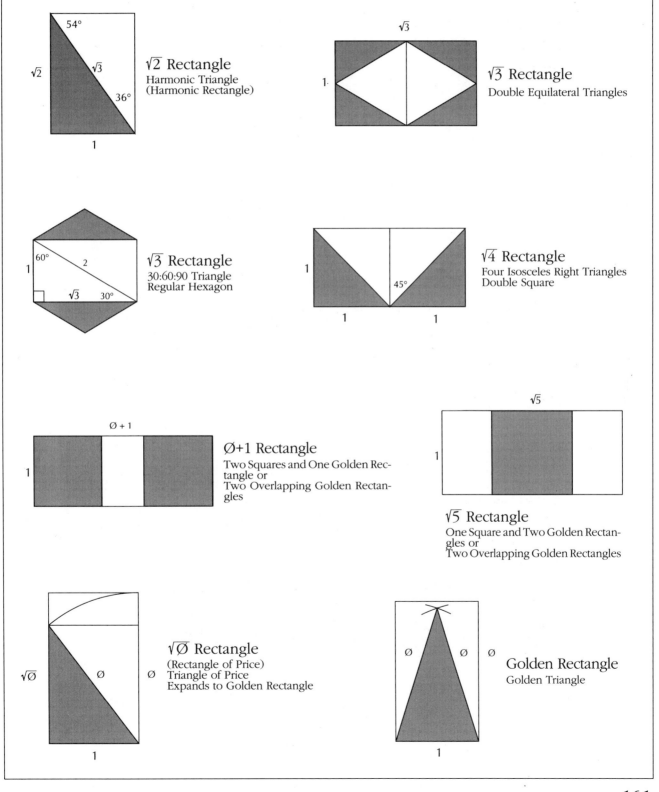

√2 Rectangle
Harmonic Triangle
(Harmonic Rectangle)

√3 Rectangle
Double Equilateral Triangles

√3 Rectangle
30:60:90 Triangle
Regular Hexagon

√4 Rectangle
Four Isosceles Right Triangles
Double Square

Ø+1 Rectangle
Two Squares and One Golden Rectangle or
Two Overlapping Golden Rectangles

√5 Rectangle
One Square and Two Golden Rectangles or
Two Overlapping Golden Rectangles

√Ø Rectangle
(Rectangle of Price)
Triangle of Price
Expands to Golden Rectangle

Golden Rectangle
Golden Triangle

Problems

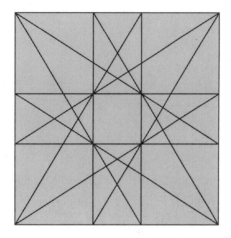

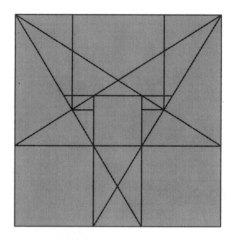

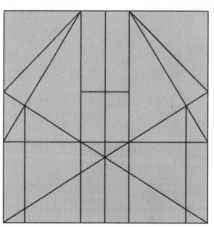

This page:
Fig. 4.44 Use with problem 3.

1 Draw a scalene triangle that is not a right triangle. Subdivide it into two triangles in such a way that one of the triangles is the gnomon of the other. *Hint: Two triangles are similar if two angles of one are congruent to two angles of the other. Remember that one of the small triangles must be similar to the larger one.*

2 Find reproductions of two paintings from different periods of Art History. Photocopy them. Using a tracing paper overlay, construct an armature for each. Compare and contrast them in a written statement.

3 Using compass and straightedge and the techniques described in this chapter, reconstruct one of the subdivisions illustrated in Fig. 4.44 using a square that measures 7" on a side.

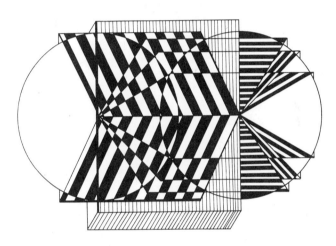

This page:
Linda Maddox. Dynamic Parallelogram variations. Technical pen and ink on paper.

4 Analyze the method of construction of *one* of the border designs illustrated on page 150 and duplicate it on a piece of paper.

5 Do an analysis of actual leaf proportions by escribing a rectangle around each leaf so that its sides are tangent to the rectangle. Do a minimum of five. Measure the sides of each rectangle and find the ratio of the length to the width. Do any of the rectangles approximate Dynamic Rectangles? If so, name them.

Hint: It might be easier to do rubbings of the leaves using light-weight bond paper and a soft pencil, *or to photocopy them and work with their images. This is not a winter problem!*

6 Construct at least six *different* √2 Parallelograms and six *different* Golden Parallelograms.

Hint: Refer to Fig. 4.26.

7 Construct a square with sides that measure 7 inches. Using your knowledge of the Golden Cut, create a harmonious subdivision of the square.

Projects

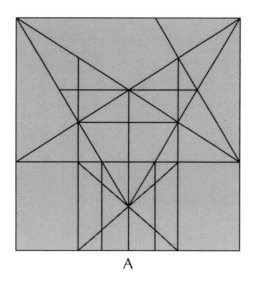

A

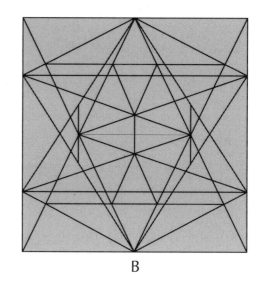

B

This page:
Fig. 4.45 Use with Project 1.

1 Deliberately changing either A or B in Fig. 4.45 as the basis for your composition, do one of the following in a minimum size of 13":

    a. a black and white variation in two dimensions

    b. a black and white variation in three dimensions

    c. a variation in value in at least 5 steps

    d. a monochromatic variation in either two or three dimensions

    e. only the primary colors plus black and white in either two or three dimensions

    f. a pair of complementary colors in either two or three dimensions.

    g. a triadic group of colors in either two or three dimensions.

2 Take the square you constructed in Problem 3 as the basis for an artwork, but change it substantially, and embellish it with your choice of material, color and technique.

3 | Using the subdivision you constructed in Problem 7 as an armature, create a collage, drawing or painting using representational subject matter (still life, portrait, landscape, etc.).

4 | Create a border design using one of the Dynamic Rectangles and the construction techniques you have learned. Take the same rectangle and turn it into a Dynamic Parallelogram. Subdivide it by the same method used for the rectangle and repeat the border design. Create a third variation of this design by skewing the parallelogram even further and using the same subdivisions. Color it in black and white, value, or hues of your choice.

5 | a. Using a 3" or 5" square, reconstruct the diagram in Fig. 4.46, given that C is the Golden Section of $\overline{AB}$, and create a variation of it in which some line segments are replaced by curves.

b. Photocopy a group of your squares and create an overall pattern from the group.

c. Color them with black and white only or use a personal color scheme.

d. Mount on a sturdy board.

6 | Based on your understanding of Northwest Coast Indian design, and the use of Dynamic Rectangles therein, create your own blanket or box design study.

7 | If the results of Problem 5 yield leaves that can be escribed by at least two different compatible Dynamic Rectangles, create a plane design with those leaves which incorporates the following materials and techniques: line drawing; crayon rubbing; tissue collage; or choose your own.

8 | Create a fantasy fish, leaf or bug that has internal proportions that fit within one or more of the rectangles discussed in this chapter. Incorporate your creation into a design in either two or three dimensions.

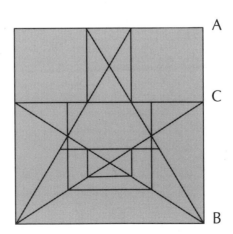

Top:
Border design using dynamic elements.

Left:
Fig. 4.46 Use with Project 5.

165

9 Using the results from Problem 4 as a resource:

    a. do a color variation of the same border design, or

    b. using multiples of the same design, do an allover pattern, or

    c. do a border that encloses an area on all four sides.

Below:
Fig. 4.47 Use with Project 11.

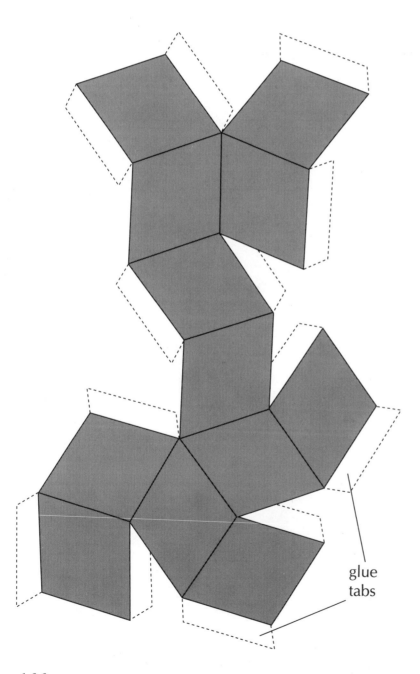

glue tabs

10 Choose any Dynamic Rectangle or Dynamic Parallelogram and construct a minimum of ten congruent figures that are large enough to manipulate comfortably. *(The photocopier is a wonderful tool for constructing congruent figures!)* Using one as a module, fold, bend or score, but do not cut, to produce a three-dimensional unit. Treat the others in the same way and then join together to form a modular sculpture.

11 Sculptural Form
The net in Fig. 4.47 shows how to lay out twelve rhombuses so that they may be folded into a three-dimensional figure called a Rhombic Dodecahedron. The appropriate rhombus is the one whose angles measure 70°32' and 109°28'. *(Hint: for your net, use a protractor and approximate the angles by measuring halfway between 70° and 71° and halfway between 109° and 110°.)* In this rhombus the ratio of the lengths of the diagonals is $\sqrt{2}$ to 1, the same as the ratio of the lengths of the diagonal to a side in the unit square. The size of the rhombus to be used is the choice of the designer, but all faces must be congruent. Therefore, it is helpful to construct one rhombus as accurately as possible and use it for a template.

While the surface is still in the plane, enhance the faces using one of the following:

    a. variations on harmonic subdivisions of the square, or

    b. divide a diagonal or side in Ø proportions and connect to points on a corresponding subdivision.

All twelve faces may be in black and white or line only, or may utilize a monochromatic, analogous or triadic color scheme.

5 Fibonacci Numbers

Once upon a time, way back in the 13th century, there lived an Italian boy named Leonardo, son of Bonaccio, later to become known as Fibonacci (1170-1250). He grew up in the city of Bugia in North Africa where his father was a customs official. As a youth he studied with Muslims who taught him the Arabic system of numerals which he realized was much less cumbersome than Roman notation. He wrote several books on geometry and trigonometry, but he is most well known for the book that he wrote at age 27 entitled *Liber Abaci*, The Book of Abacus, which became the prime vehicle for introducing Hindu-Arabic numbers to the European world. He is considered the greatest mathematician of the Middle Ages.

It was, however, a particular theoretical problem in *Liber Abaci* that has kept his name alive these many centuries. His solution to this problem offered a special sequence of numbers that was to have implications in other fields, beyond mathematics, such as art, botany, biology, and music. Even now, mathematicians continue to explore these special numbers.

First, let us look at the problem he set for himself, then let us examine the conclusions. Keep in mind that this problem is more idealistic than realistic in terms of natural animal reproduction.

This page:
Leonardo Fibonacci (1170 - 1250). Mathematician of the Middle Ages.

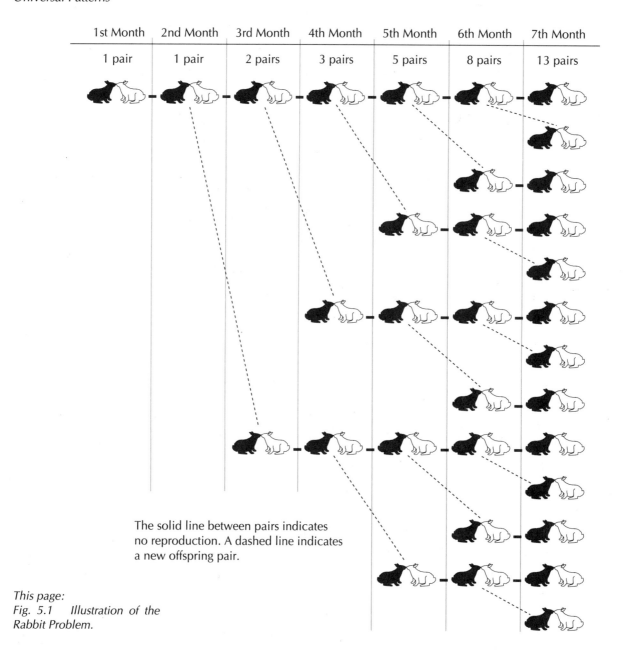

1st Month	2nd Month	3rd Month	4th Month	5th Month	6th Month	7th Month
1 pair	1 pair	2 pairs	3 pairs	5 pairs	8 pairs	13 pairs

The solid line between pairs indicates no reproduction. A dashed line indicates a new offspring pair.

This page:
Fig. 5.1   Illustration of the Rabbit Problem.

The Rabbit Problem

How many rabbits would you have at the beginning of each month if you start with a single pair and:

a. each adult pair gives birth to a new pair every month,

b. each pair reproduces from the second month on, and

c. no rabbit dies.

You will notice that a pattern emerges giving rise to a numerical **sequence** whose terms are 1, 1, 2, 3, 5, 8, 13. If we continue watching these rabbits (and none gets stewed!) the pattern would continue as 1, 1, 2, 3, 5, 8, 13, 21, 34, 55,.... Each term is the sum of the two preceding terms. In honor of the man who set the problem, it has come to be known as the Fibonacci sequence.

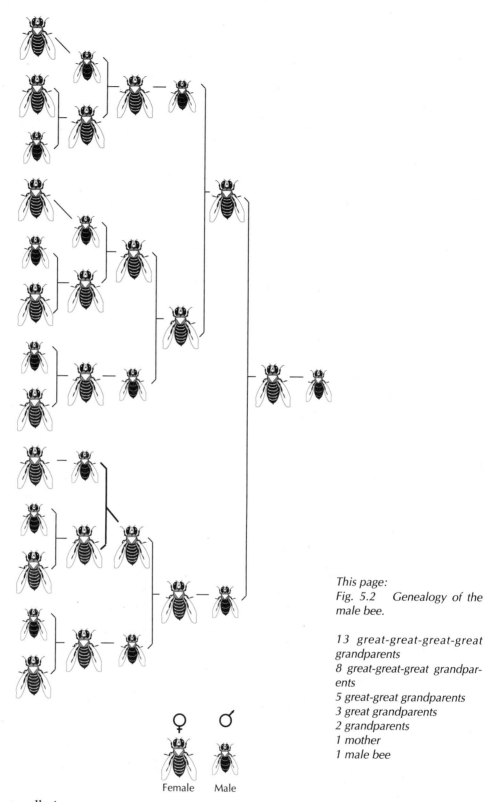

This page:
Fig. 5.2   Genealogy of the male bee.

13 great-great-great-great grandparents
8 great-great-great grandparents
5 great-great grandparents
3 great grandparents
2 grandparents
1 mother
1 male bee

♀   ♂

Female   Male

This sequence occurs naturally in the genealogy of the male bee. He has a single parent, only a mother, whereas a female bee has both a mother and a father. We trace the ancestry of the male in Fig. 5.2.

If we observe the flowers, leaves and branches of the *Arcillea ptarmica* (sneezewort), we find several occurrences of Fibonacci numbers. The simplified diagram in Fig. 5.3 illustrates the manner in which the branch and leaf patterns yield the Fibonacci sequence. In each case, consider the number of leaves or branches cut by various horizontal planes. In the field, Nature does not follow this pattern absolutely, but the tendency for these numbers to appear in natural branching patterns is too great to discount.

Below:
Fig. 5.3 Fibonacci elements in the sneezewort plant.

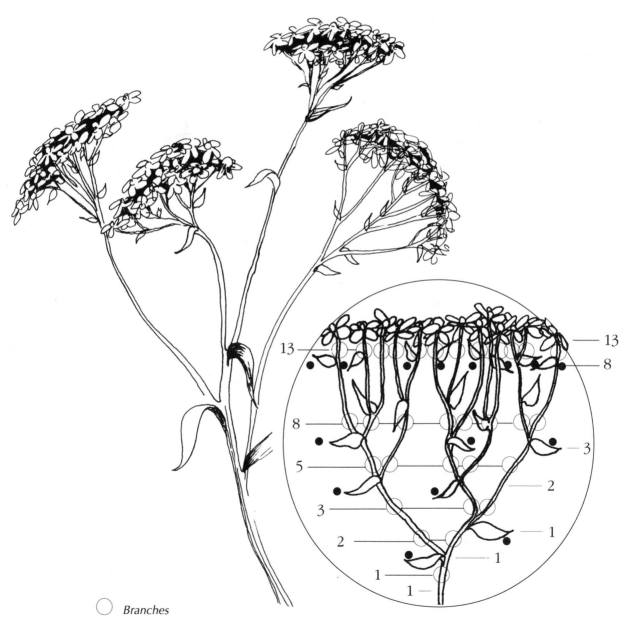

◯　**Branches**

●　**Leaves and flowers**

This page:
Ruth Des Roches. Student
Work. Chapter 5 Project 2.
Mixed media.

Mathematical Properties of the Fibonacci Numbers

The Fibonacci sequence has many special properties that have intrigued mathematicians throughout the years and continue to do so. We will examine a few of them here.

The chart on the next page illustrates how quickly the numbers get large. F_{50} has 11 digits. Obviously, generating the terms by hand beyond this point is cumbersome and time consuming. However, the process is simplified radically by the use of the computer, and mathematicians have been able to examine the larger terms with considerable ease. For example, F_{19137} has been computed and was found to be 4,000 digits long.

We have already noted that each term in the sequence is the sum of the two preceding terms. There are several other sums of interest as well.

The First Fifty Fibonacci Numbers

n indicates the position of the term in the Fibonacci Sequence.

F_n indicates the Fibonacci number in the *nth* position. For example: the 32nd Fibonacci number is 2,178,309.

Notice that beyond F_{15} *the ratio of any consecutive pair of Fibonacci numbers is Ø, correct to at least five decimal places.*

$$\frac{F_4}{F_3} = \frac{3}{2} = 1.50000$$

$$\frac{F_8}{F_7} = \frac{21}{13} \approx 1.61538$$

$$\frac{F_{11}}{F_{10}} = \frac{89}{55} \approx 1.61818$$

$$\frac{F_{13}}{F_{12}} = \frac{233}{144} \approx 1.61806$$

$$\frac{F_{15}}{F_{14}} = \frac{610}{377} \approx 1.61804$$

$$\frac{F_{16}}{F_{15}} = \frac{987}{610} \approx 1.61803$$

n	F_n	n	F_n
1	1	26	121,393
2	1	27	196,418
3	2	28	317,811
4	3	29	514,229
5	5	30	832,040
6	8	31	1,346,269
7	13	32	2,178,309
8	21	33	3,524,578
9	34	34	5,702,887
10	55	35	9,227,465
11	89	36	14,930,352
12	144	37	24,157,817
13	233	38	39,088,169
14	377	39	63,245,986
15	610	40	102,334,155
16	987	41	165,580,141
17	1,597	42	267,914,296
18	2,584	43	433,494,437
19	4,181	44	701,408,733
20	6,765	45	1,134,903,170
21	10,946	46	1,836,311,903
22	17,711	47	2,971,215,073
23	28,657	48	4,807,526,976
24	46,368	49	7,778,742,049
25	75,025	50	12,586,269,025

Some of the interesting properties of the Fibonacci sequence are:

$1 + 1 + 2 = 4$, and $4 = \boxed{5} - 1$.

$1 + 1 + 2 + 3 = 7$, and $7 = \boxed{8} - 1$.

$1 + 1 + 2 + 3 + 5 = 12$, and
$\quad 12 = \boxed{13} - 1$.

$1 + 1 + 2 + 3 + 5 + 8 = 20$, and
$\quad 20 = \boxed{21} - 1$.

$1 + 1 + 2 + 3 + 5 + 8 + 13 = 33$, and
$\quad 33 = \boxed{34} - 1$.

Notice that in each case the sum of the terms is a number that is one less than another Fibonacci number. In fact, if the first n terms (where n is any counting number) are added, i.e. $F_1 + F_2 + F_3 + ... + F_n$, then the sum is one less than F_{n+2} (the second term after F_n).

In general,
$$F_1 + F_2 + F_3 + ... + F_n = F_{n+2} - 1.$$

Now, let us see what happens when we add only the *even numbered* terms or only the *odd numbered* terms.

The *even numbered* terms are:
$\quad F_2, F_4, F_6, F_8, F_{10}, ...$
$\quad$ or, 1, 3, 8, 21, 55, ...

$1 + 3 = 4$, and $4 = \boxed{5} - 1$.

$1 + 3 + 8 = 12$, and $12 = \boxed{13} - 1$.

$1 + 3 + 8 + 21 = 33$, and $33 = \boxed{34} - 1$.

$1 + 3 + 8 + 21 + 55 = 88$, and
$\quad 88 = \boxed{89} - 1$.

In general,
$F_2 + F_4 + F_6 + ... + F_{2n} = F_{2n+1} - 1$, or the sum is one less than the next term in the sequence.

The *odd numbered* terms are:
$\quad F_1, F_3, F_5, F_7, F_9, ...$
$\quad$ or 1, 2, 5, 13, 34, ...

$1 + 2 = \boxed{3}$.

$1 + 2 + 5 = \boxed{8}$.

$1 + 2 + 5 + 13 = \boxed{21}$.

$1 + 2 + 5 + 13 + 34 = \boxed{55}$.

In general,
$F_1 + F_3 + F_5 + ... + F_{2n-1} = F_{2n}$, or the sum is the next term in the sequence.

Another pattern that is evident in this sequence is that the *square* of any term differs from the *product* of the terms on either side of it by ± 1. For example, consider three consecutive terms such as 8, 13, and 21.
$\quad 13^2 = 169$
$\quad (8)(21) = 168$
$\quad$ and $13^2 - (8)(21) = 1$.

If we had chosen 5, 8, and 13 we would have:
$\quad 8^2 = 64$
$\quad (5)(13) = 65$
$\quad$ or $8^2 - (5)(13) = -1$

In general,
$$(F_n)^2 - (F_{n-1})(F_{n+1}) = \pm 1$$

The aforementioned properties are but a few of those that set the Fibonacci Sequence apart from others. For the mathematically curious reader, these properties are proved in Appendix B.

changing the divisor and dividend to compute the reciprocal of Ø. With the exception of the number 1 before the decimal, the two decimal representations were identical up to the 4,598th place.

In addition to this direct connection to Ø, we can generate another sequence that relates both to the Fibonacci numbers and to Ø. The **Golden Sequence** is the summation sequence whose initial terms are 1 and Ø.

$$1, \varnothing, 1 + \varnothing, 1 + 2\varnothing, 2 + 3\varnothing, 3 + 5\varnothing, 5 + 8\varnothing, 8 + 13\varnothing, \dots$$

Fibonacci and Phi

We have found the most interesting property of the Fibonacci Sequence to be the one that connects it to Ø. Notice in the table of Fibonacci numbers that as n gets larger, the ratio of any pair of consecutive terms in the sequence gets closer to Ø.

The computer has enabled us to examine Ø to many decimal places. In an article for the *Fibonacci Quarterly*, Murray Berg describes the calculation of Ø to 4,600 decimal places by dividing F_{11004} by F_{11003}, each of which contains 2,300 digits. Obviously, this goes beyond hand calculation as it would be more than slightly inconvenient to divide a 2,300 digit number by another 2,300 digit number and carry out the division to 4,600 places. This particular exercise was completed in 20 minutes on an IBM 1401 computer. The calculation was checked by inter-

Notice the appearance of a double Fibonacci Sequence. Beginning with the third term, the constants are terms in the Fibonacci Sequence. Beginning with the second term (since Ø = 1Ø), the **coefficients** of Ø are in Fibonacci sequence. Further examination reveals the sequence to be one we have seen before.

Remembering that Ø is the solution to the equation $x^2 - x - 1 = 0$, we have $\varnothing^2 - \varnothing - 1 = 0$, or $\varnothing^2 = 1 + \varnothing$. We can rewrite the first three terms of the sequence 1, Ø, $\varnothing^2$. Since this is a summation sequence, and we know from Chapter 2 that $\varnothing^n + \varnothing^{n+1} = \varnothing^{n+2}$, we can rewrite the entire sequence:

$$1, \varnothing, \varnothing^2, \varnothing^3, \varnothing^4, \dots.$$

That is, the Golden Sequence is the sequence of the powers of Ø that we examined in Chapter 2.

Fibonacci and Design

Fibonacci numbers translate in a practical way into dimensions of length and width important for the visual artist. Despite the fact that the ratio of two consecutive Fibonacci numbers rapidly approaches Ø as *n* increases, using a pair toward the beginning of the sequence offers a tool for constructing a rectangle that approximates the Golden Rectangle, but with sides whose lengths are commensurable. *(Remember that the Dynamic and Ø-Family Rectangles have sides whose lengths are incommensurable.)*

Let us examine another way in which the Fibonacci Sequence is beneficial to the designer. It provides a way to harmonically divide line segments and bounded portions of the plane in a manner such as illustrated in Fig. 5.5. The sides of the rectangle have been divided into lengths that are in Fibonacci sequence, beginning at the midpoints and working toward the endpoints. The points of division on one side are then connected by line segments to the corresponding points on the opposite side. The design is completed by darkening alternate areas. The construction for this type of division follows on the next page.

Center left and right:
Fig. 5.4 Approximations of Golden Rectangles using sides whose lengths are commensurable Fibonacci numbers. Length to width ratios are 1.6 and 1.625.

Bottom:
Fig. 5.5 Fibonacci subdivision of a rectangle.

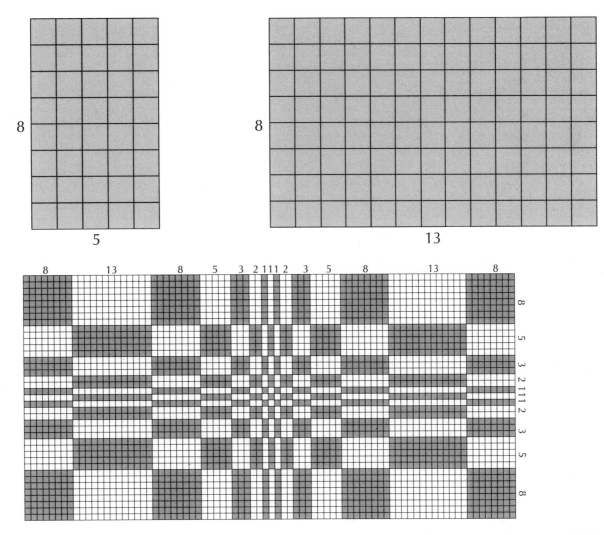

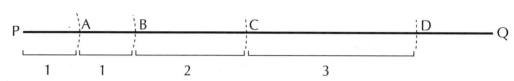

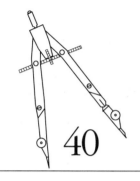

40

Divide a Line Segment into Lengths Corresponding to a Sequence of Fibonacci Numbers

Given $\overline{PQ}$

1. Choose a unit length, and mark off units in sequence from P toward Q in the following manner.

2. Set the compass to measure the unit length and cut an arc on $\overline{PQ}$. Label the point of intersection A (PA = 1).

3. Without changing the setting, set the metal tip on A and cut another arc on $\overline{PQ}$. Label the point of intersection B (AB = 1).

4. Open the compass to measure PB and, with the metal tip on B, cut another arc on $\overline{PQ}$. Label the point of intersection C (BC = 2).

5. Open the compass to measure AC, place the metal tip on C, and cut another arc on $\overline{PQ}$. Label the point of intersection D (CD = 3).

6. Repeat, each time setting the compass to measure the previously constructed consecutive pair of segments. Extend as far as desired.

Now $\overline{PQ}$ is divided into lengths corre-sponding to Fibonacci numbers.

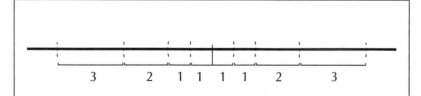

If the subdivision begins about the midpoint, a palindromic sequence of units is achieved. This is the method of subdivision used in Fig. 5.5.

Fibonacci divisions are not limited to rectangles. In fact, as the number of sides of the polygon increases, so do the number of design possibilities (see Fig. 5.6). They become almost endless with the introduction of color, curves, and sequential divisions of line segments beginning at points other than the midpoint. If multiple units of a design are required for a project, a photocopier is very helpful.

This page:
Fig. 5.6 Fibonacci subdivisions of several different polygons. Ann MacLean. Technical pen and ink on paper.

Facing pages:
Student works. Chapter 5
Project 1.
Top left: Lana Yonkers
Bottom left: Joan Bibaud
Far right: Sandy Sedler

Above:
Spiral phyllotaxis in the head of a daisy.

Fibonacci and Phyllotaxis

In the natural world the arrangement of leaves, buds, and seeds is termed phyllotaxis. Its connection with Ø and Fibonacci is evident in the phenomenon of gnomonic growth in which the shape remains the same as the size changes. We saw this in Chapter 2 when we created successively larger Golden Rec tangles.

In organic forms that grow from the center outward, a spiral formation occurs because the spiral fills space efficiently in a perfectly regular manner and is capable of infinite expansion. The most striking example of spiral phyllotaxis occurs in the head of the common sunflower. The rhombic seed pockets form a series of intersecting spirals in a plane. Under normal circumstances of growth these almost always provide a constant ratio of 55 longer curves travelling in one direction intersecting 89 shorter ones travelling in the other direction. Some heads may be found to have 89 longer curves inter-secting 144 shorter ones.

This page:
This drawing illustrates the fact that the number of petals on particular flowers is frequently a Fibonacci number.

Number of petals	Type of flower	
3	irises, lilies	Pine needles tend to
5	buttercups, primroses	grow in clusters of 2,
8	some delphiniums	3, or 5 depending on
13	ragwort, some marigolds	the species.
21	some asters, marigolds	
34, 55, 89	different types of daisies	

If we examine these ratios closely we see that both 89/55 and 144/89 give close approximations to Ø, and that 55, 89, and 144 are, in fact, all terms in the Fibonacci Sequence. Like the sunflower, the head of a daisy has two sets of intersecting spirals, which utilize another pair of Fibonacci numbers with a ratio approximating Ø. Usually there are 21 seed spirals in one direction intersecting 34 seed spirals in the other.

Fibonacci spirals into the third dimension in the phyllotaxis of pineapples and the cones of evergreen trees. The pineapple is more mathematically analogous to the cylinder while evergreen cones are more analogous to conical shapes. The hexagonal scaly plates of the pineapple are arranged in helical curves. The seed pockets on the cone, however, form different kinds of curves called concho spirals which come closer to approximating logarithmic spirals. We shall examine both of these types of curves in detail in the next chapter.

Apart from the shape of the curves, the spirals still appear in Fibonacci numbers. If we consider the pineapple as a cylinder, cut it vertically and flatten it, we can see the spirals transformed into straight lines in the plane (see Fig. 5.8). In three dimensions (Fig. 5.7), we can see how the straight lines from the illustration of the flattened section wrap around the fruit.

Top:
Fig. 5.7 Spirals around the pineapple.

Bottom:
Fig. 5.8 Planar view of the surface of a pineapple. The vertical lines represent the cut, and the portion between them shows the flattened scales on the skin. There are 5 sharp curves to the right, 13 steep curves, again to the right, as well as 8 more gently sloping curves to the left.

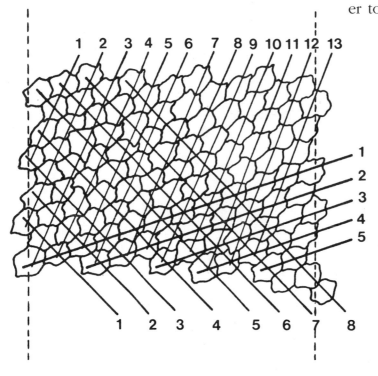

In the cones of evergreen trees, the seed-protecting scales form two groups of intersecting spirals. These frequently appear in the ratios 8:5 or 13:8. In addition to the fact that the number of spirals in either direction is a Fibonacci number, it is often the case that the number of seeds in any one spiral is also a Fibonacci number. The ratios of the groups of curves in spiral phyllotaxis seem frequently to come from members of the following sequence:

$$\frac{1}{1}, \frac{2}{1}, \frac{3}{2}, \frac{5}{3}, \frac{8}{5}, \frac{13}{8}, \frac{21}{13}, \frac{34}{21}, \frac{55}{34}, \frac{89}{55}, \frac{144}{89}, \dots$$

In this sequence, each term is composed of a pair of consecutive Fibonacci numbers such that, with the exception of the initial term, the sequence formed by the numerators and the one formed by the denominators are both Fibonacci Sequences. As the terms progress the ratio comes closer and closer to Ø.

Turning from the arrangement of seeds to the arrangement of leaves around a stem, we again encounter a phenomenon that suggests a sequence of Fibonacci numbers. The leaves of a majority of higher plants are arranged in a particular spiral sequence growing up and out from the cylindrical stem. If we choose a particular leaf on a stem and travel up the stem passing through consecutive leaf nodes, we will eventually arrive at a leaf situated directly above the first one we chose. If t is the number of complete turns made around the stem, and n the number of leaves passed along the way, then the fraction $\frac{t}{n}$ is said to be the **divergency constant** for that particular species.

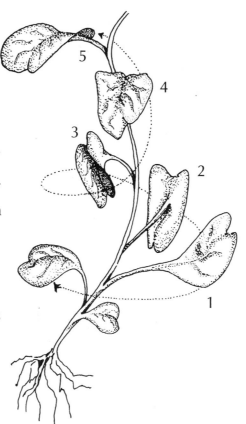

Top:
Evidence of spiral structure in evergreen cones.

Bottom:
Fig. 5.9   Leaf number five is situated directly above the un-numbered leaf. When moving from one to the other, five leaves are passed through in two complete turns around the stem. Thus, the divergency constant for this type of plant is 2/5.

185

Divergencies are generally found to be terms from the following sequence:

$$\frac{1}{2}, \frac{1}{3}, \frac{2}{5}, \frac{3}{8}, \frac{5}{13}, \frac{8}{21}, \frac{13}{34}, \frac{21}{55}, \dots$$

Again, the numerators and denominators are in Fibonacci sequence. This time, however, the individual terms are composed of fractions of the form $\frac{F_n}{F_{n+2}}$. The divergency constants for some specific types of trees are noted on this page.

Nature uses this arrangement to provide the proper amount of sunlight for the plant. This way of distributing leaves seems to give the necessary exposure to sunlight as well as protection when needed. Under perfect conditions, which the mathematician can create but Nature rarely, if ever, does, a plant will obtain maximum exposure to sunlight and minimum overlapping of leaves if each leaf is separated from the succeeding one by an angle of about 137.5°. At this angle, provided the shoot be kept perfectly straight, no two leaves would ever be exactly one over the other. This angle is called the **Ideal Angle** or sometimes the **Fibonacci Angle** and bears a close relationship to Ø (see Fig. 5.10). Maximum exposure to sunlight is not always a healthy condition for a plant. Since the needs of plants vary, the Ideal Angle of divergence would not provide optimal growing conditions for every species, hence the variety of divergency constants.

Divergencies for some specific types of trees:

elm	1/2
beech, hazel	1/3
apple, oak, apricot, and poplar	2/5
pear, weeping willow	3/8
willow, almond, and pussy willow	5/13

Top:
Spiral phyllotaxis in the branches of trees.

Right:
Fig. 5.10   If we multiply the total number of degrees in a circle, 360°, by the fraction 1/Ø (the reciprocal of Ø), we get an angle of about 222.5°. The angle needed to complete the circle measures 137.5°, the Ideal Angle.

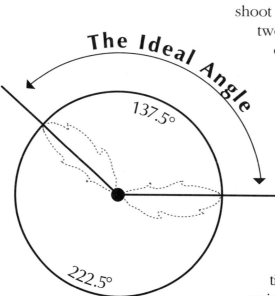

The Ideal Angle

137.5°

222.5°

Fibonacci and Musical Composition

We are greatly indebted to J. Michael Pope, one of our former students, who is a practicing composer and musician. Exposure to the Fibonacci Sequence in our course sparked a desire to utilize these ideas in his own work with musical form. He sees great potential in the Fibonacci Sequence as a musical structure. An example of his work with explanation follows.

A composer of music may use the Fibonacci sequence to proportionally segment the dimension of his or her art form: time. For example, in determining a skeletal design of the duration of the movements of a symphony, the composer may utilize the numbers of the sequence to create time frames that are in Ø proportion to each other. The duration of individual sections of each movement may also be divided proportionally using the Fibonacci sequence. Or, a composer may use the sequence to produce rhythms within the individual sections. The excerpt from the chamber work, Yes Yes Men Live, for two violins and bassoon, utilizes the Fibonacci sequence to generate rhythmic change.

At this point in the piece, the composer creates a rhythmic pattern that gradually slows (ritardando) the melodic alternation. At its longest duration the melodic material reverses and the rhythmic pattern gradually increases the speed of alternation (accelerando) in proportion to the slowing. A reduction of the passage to a single line displays the use of the Fibonacci sequence.

J. Michael Pope

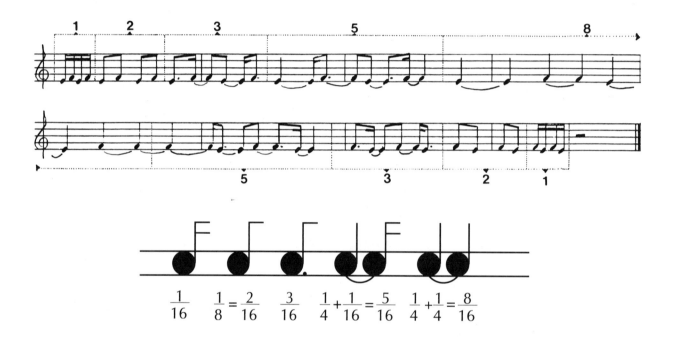

$$\frac{1}{16} \qquad \frac{1}{8}=\frac{2}{16} \qquad \frac{3}{16} \qquad \frac{1}{4}+\frac{1}{16}=\frac{5}{16} \qquad \frac{1}{4}+\frac{1}{4}=\frac{8}{16}$$

Top and center:
Fig. 5.11 Fibonacci numbers in relation to musical time. The key to understanding this structure is to know that the sixteenth note is considered the unit measure. Notice that the numerators are in Fibonacci progression.

Bottom:
Student work. J. Michael Pope. Chapter 5 Project 1.

This page:
J. Michael Pope
Top:
Vamp #1 Minimal.
Bottom:
Vamp #2 Mindscape.
Visual expressions of a musical
score. Pen and ink on paper.

Visual and Aural Musical Perception

Michael has translated his musical composition into the following visual variations because of his feelings that we can "hear with our eyes and see with our ears". In each case the dimensions of the rectangle are in a ratio of 3 to 2, and the subdivision of the interior space is also in the Fibonacci progression. Beginning on the far left and moving from top to bottom, the three horizontal bands represent the instrumental parts of the two violins and the bassoon respectively. The central portion in each depicts, for him, the fusion of all three parts. Vamp #1 is his *Minimal* interpretation, while Vamp #2 he considers a *Mindscape.*

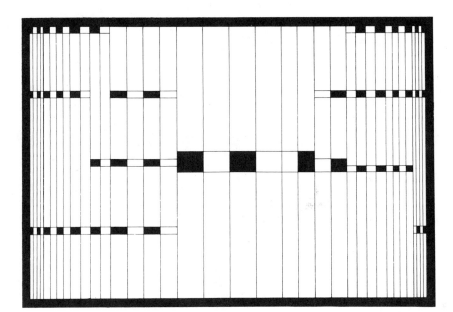

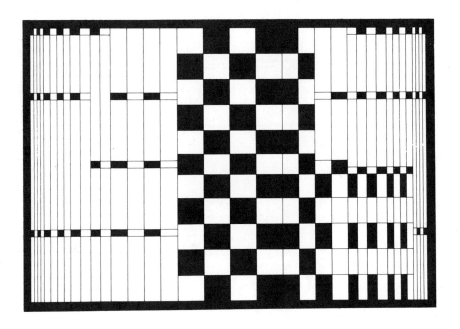

This page:
Student work. Claire Melanson.
Chapter 5   Project 4. Strips
from found images woven to-
gether.

Fibonacci, Phi and Phood

In our research, we have found Fibonacci numbers springing up in strange places. Some of our students provided us with these recipes, and we feel sure that you will agree that the Ø connection makes them especially wonderful. We share them with you for your gustatory delight.

Fibonacci Fudge
Sharon Maniates

This recipe was invented especially for this chapter. Notice that there are *8* ingredients combined according to directions containing *8* sentences.

 5 c. granulated sugar
13 oz. evaporated milk
 1 stick of butter
13 oz. marshmallow (fluff type)
 1 tsp. salt
 1 tsp. vanilla
 2 c. walnuts
 2-12 oz. packages of semi-sweet chocolate bits

Combine the first *5* ingredients in a large pan. Stir over medium heat until well blended. Boil for *5* minutes at medium-high heat. Overcook rather than undercook. Remove from heat and stir in the last *3* ingredients until chocolate is melted and well blended. Pour into *2* buttered *8"* x *8"* pans. Allow to cool and cut into squares.
Makes approximately *5* pounds.

Fibonacci Baked Beans
Robin Johnson

1 can (15.5 oz.) kidney beans, drained
1 can (15.5 oz.) butter beans, drained
2 cans (21 oz.) pork and beans
3 chopped onions
5 miscellaneous ingredients
    1 tsp. garlic powder
    $\frac{1}{2}$ tsp. dry mustard
    $\frac{3}{4}$ c. brown sugar
    $\frac{1}{2}$ c. cider vinegar
    $\frac{1}{4}$ c. catsup
8 slices bacon, diced

Heat oven to 350°. Saute onion and bacon. Drain and mix with all the other ingredients. Pour into casserole and bake 60 - 70 minutes, or until hot.

Fibonacci Connections

LOBSTERS: Grow in size only when they discard their shells. In the first year this occurs 8 times. In the second year this occurs 5 times. In the third year this occurs 3 times. After that, the male sheds 2 times per year while the female sheds only 1 time per year.

Fibonacci Baklava
Harriet Lincoln

Pastry

$1\frac{1}{2}$ lbs. Filo dough (found in the refrigerator case at the supermarket)

$\frac{1}{2}$ c. sugar

1 lb. butter, melted

3 c. walnuts, fine chopped (may be done in a blender)

1 tsp. cinnamon

$\frac{1}{2}$ lb. almonds, finely chopped

2 zweiback crackers, ground

1 Tbs. water

Mix together the cinnamon, sugar, zweiback and chopped nuts.
Brush bottom of 9" x 13" pan with melted butter.
Place *2* Filo sheets on bottom of pan. Brush with melted butter. Add *3* more Filo sheets. Layer with some of the nut mixture. Follow with *3* buttered Filo sheets alternating with nut mixture *5* more times, or, until all but *8* filo sheets are used. Finally add the last *8* buttered sheets. Sprinkle with water.

Cut into diamond shapes. Bake at 400°for 15 minutes, 350° for 30 min-utes, and 300° for one hour until golden color. Place pan on wire rack to cool. Make syrup from the direc tions given here while the pastry bakes.

Honey Syrup
2 whole cloves
1 c. honey
1 Tbs. brandy (optional)
1 small lemon
1 c. sugar
1 c. water
1 2" piece of cinnamon stick

Remove rind from lemon. Squeeze out $1\frac{1}{2}$ tsp. of juice for reserve. Place lemon rind, sugar, water, cinnamon stick, cloves in a saucepan. Bring to boil. Reduce heat and cook 25 min-utes without stirring. It will be syrupy. Stir in honey. Stir in reserve lemon juice and brandy.

Pour cooled syrup over cooled pastry. Let stand overnight and then serve.

This page:
Peter Carey. Student Work.
Chapter 5   Project 4. Rubber
stamp and cut paper.

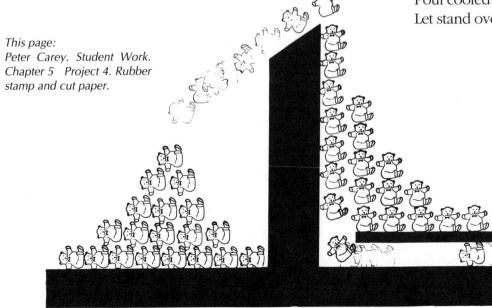

Problems

Above:
Natural forms exhibiting spiral phyllotaxis.

[1] Figure the divergency constant for 5 different house plants (see Fig. 5.9). Develop a chart that gives the names of the plants together with their divergency constants. You may include drawings, photographs or photocopies.

[2] Draw 3 line segments each at least 8 inches long. Divide each one palindromically into Fibonacci lengths:

    a. starting at the midpoint,

    b. starting at the endpoints, and

    c. starting at the two points that divide the segment into thirds. *(You will notice an irregularity at the midpoint in this case.)*

[3] Using cardboard, wood or plastic, devise a Fibonacci ruler to use for design purposes. Consider the placement of the initial term and size of unit for maximum effectiveness.

[4] Divide the sides of the following shapes into lengths corresponding to palindromic sequences of Fibonacci numbers and connect the points in a manner of your choosing.

    a. Pentagon,

    b. any of the Dynamic Rectangles or Dynamic Parallelograms,

    c. any of the Special Triangles,

    d. any member of the Phi-Family Tree.

Projects

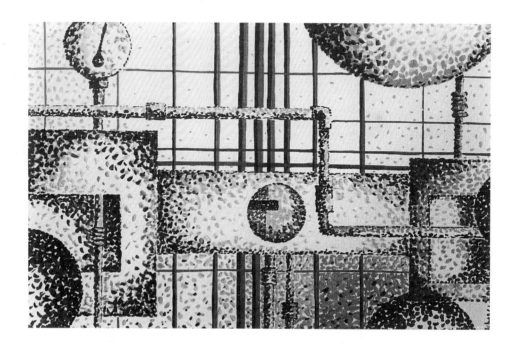

This page:
Lisa Petrucci. Student Work.
Chapter 5 Project 1. Gouache
on paper.

1 Using one or more of the results of problem 4, create a two-dimensional composition or a three-dimensional object.

2 Using the concept of the divergency constant in the arrangement of leaves around the stem of a plant, create a representational or abstract sculpture.

3 Using found plant or leaf images from magazines or photocopied from books, create a collage using Fibonacci numbers.

4 Use a combination of Fibonacci numbers as the underlying structure for the creation of a two or three-dimensional artwork.

5 Develop a design for a striped or plaid fabric using Fibonacci numbers and a limited palette of color.

6 Create a piece of Fibonacci Finery using a Fibonacci number of beads, buttons, feathers, ornaments or found objects.

7 Develop a recipe for a snack, salad or breakfast cereal that uses Fibonacci numbers. Make a batch to share with your friends. Send the recipe to us!

Further Reading

Bachmann, T. and P. Bachmann. "An Analysis of Bela Bartok's Music through Fibonaccian Numbers and the Golden Mean". *Music Quarterly*. (1979): pp. 72 - 82 .

Bicknell, Marjorie and Verner E. Hoggatt, Jr. Ed. *A Primer for the Fibonacci Numbers*. San Jose: The Fibonacci Association, 1973.

Cook, Theodore Andrea. *The Curves of Life*. New York: Dover Publications, Inc., 1979.

Garland, Trudi Hammel. *Fascinating Fibonaccis*. Palo Alto: Dale Seymour Publications, 1987.

Huntley, H.E. *The Divine Proportion*. New York: Dover Publications, Inc., 1970.

Jacobs, Harold R. *Mathematics A Human Endeavor*. New York: W.H. Freeman and Company, 1982.

Jean, Roger V. *Mathematical Approach to Pattern and Form in Plant Growth*. New York: John Wiley & Sons, 1984.

Stokes, William. *Notable Numbers*. Palo Alto: William Stokes, 1974.

6 Spirals

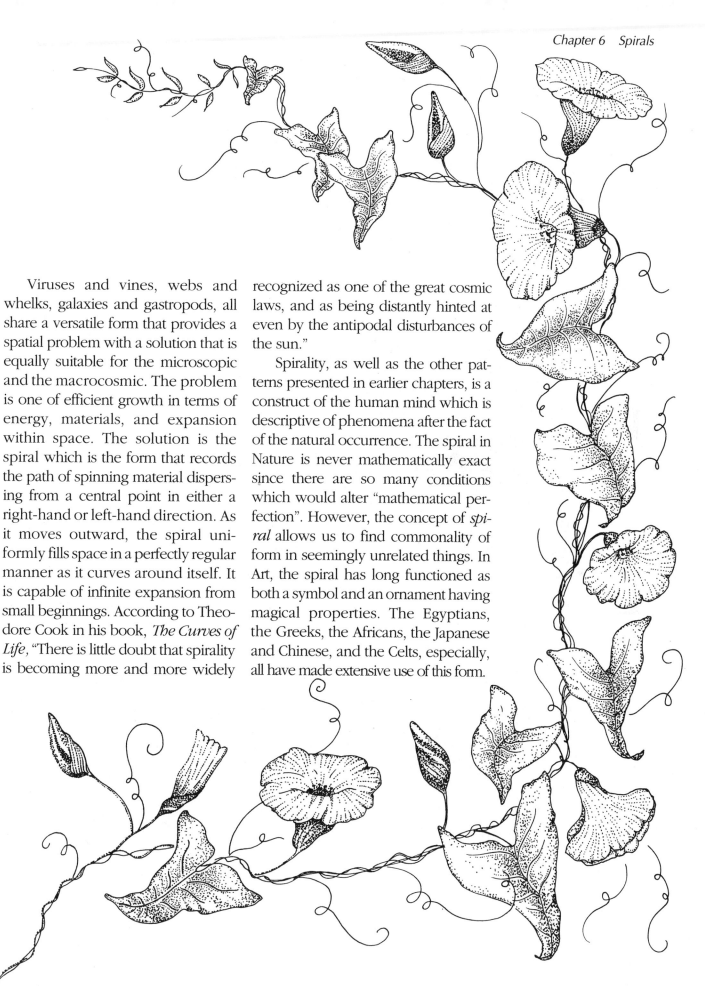

Viruses and vines, webs and whelks, galaxies and gastropods, all share a versatile form that provides a spatial problem with a solution that is equally suitable for the microscopic and the macrocosmic. The problem is one of efficient growth in terms of energy, materials, and expansion within space. The solution is the spiral which is the form that records the path of spinning material dispersing from a central point in either a right-hand or left-hand direction. As it moves outward, the spiral uniformly fills space in a perfectly regular manner as it curves around itself. It is capable of infinite expansion from small beginnings. According to Theodore Cook in his book, *The Curves of Life,* "There is little doubt that spirality is becoming more and more widely recognized as one of the great cosmic laws, and as being distantly hinted at even by the antipodal disturbances of the sun."

Spirality, as well as the other patterns presented in earlier chapters, is a construct of the human mind which is descriptive of phenomena after the fact of the natural occurrence. The spiral in Nature is never mathematically exact since there are so many conditions which would alter "mathematical perfection". However, the concept of *spiral* allows us to find commonality of form in seemingly unrelated things. In Art, the spiral has long functioned as both a symbol and an ornament having magical properties. The Egyptians, the Greeks, the Africans, the Japanese and Chinese, and the Celts, especially, all have made extensive use of this form.

The Archimedean Spiral

The Greek philosopher and mathematician Archimedes lived in the third century B.C. He attended the university at Alexandria where Euclid had taught. He returned home to Syracuse where he spent his time writing, solving problems of geometry, and inventing mechanical devices which were used to defend the city when under attack by the Romans. Despite the cleverness of his designs, the Romans eventually took both the city and his life.

Although the political leaders sought him out for his solutions to practical problems, primarily military, he was not himself a political creature, but rather was a thinker. Three works of Archimedes on plane geometry, one of which is on spirals, have survived the intervening centuries.

The Archimedean spiral is the simplest one. It is the spiral exemplified by the grooves in a phonograph record (or for the modern generation, the coil of a spool of audio or video tape), or a coiled rope. It depends, in the real world, upon the uniform thickness of the material coiled as well as the uniform tightness of the coil. Fig 6.1 visually explains this curve.

The spiral has specific mathematical properties. It is a curve traced by a point continually moving around and away from a fixed point, called the **pole**. The manner in which the point moves away from the pole determines the particular type of spiral.

Above:
Archimedean Spiral in basketry.

Concentric circles provide a framework for the spiral which moves at a uniform rate away from the pole. Any spiral which lengthens at a constant rate per constant turn is an Archimedean spiral. When these two constants are determined, construction of the spiral can be accomplished fairly easily by the following method (Construction 41). Note that this "construction" differs from the previous ones in that the required tools are other than compass and straightedge.

> The spiral form is connected to the concept of the maze and the labyrinth, which is in turn connected to the secrets of the earth.

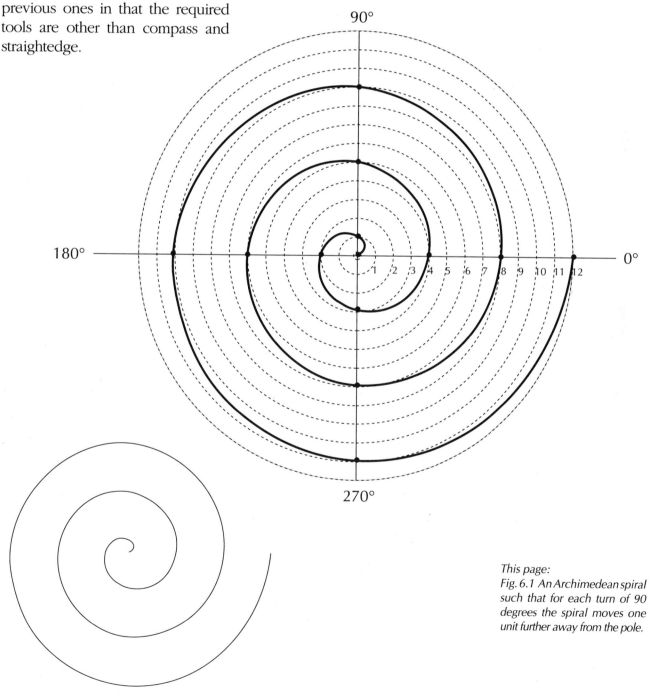

This page:
Fig. 6.1 An Archimedean spiral such that for each turn of 90 degrees the spiral moves one unit further away from the pole.

199

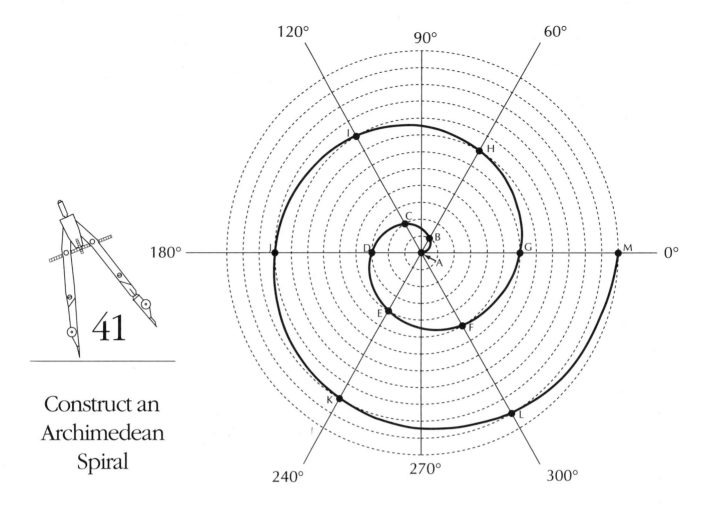

41

Construct an
Archimedean
Spiral

Given the rate at which the spiral
moves away from the pole per constant turn, in this case one unit per
turn of 60°

1. Use a sheet of commercial polar
graph paper or construct concentric
circles such that the radii increase by
a constant rate.

2. Locate the pole and label it A.

3. Locate point B at the intersection of
the first circle and the radius vector
corresponding to 60°.

4. Locate point C at the intersection
of the next larger circle and the
radius vector corresponding to 120°
(60° + 60°).

5. Continue the process, each time
moving to the next larger circle for
each subsequent turn of 60° to get
points D, E, F, G, H, I, J, K, L, and M.

6. Join the points in order with a
smooth curve. (A commercially made
French curve is a helpful tool here.)

Now the curve is an Archimedean
spiral.

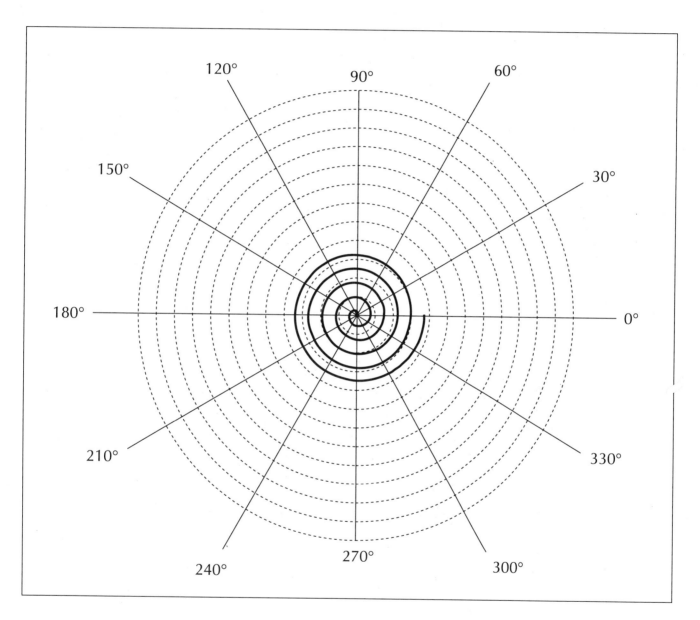

Above:
Fig. 6.2 Archimedean spiral.This one is much "tighter" since it increases only 1/6 unit per turn of 30°. Try to draw more of the spiral in the remaining space on the polar grid. The border design at the top of the page was created with this spiral.

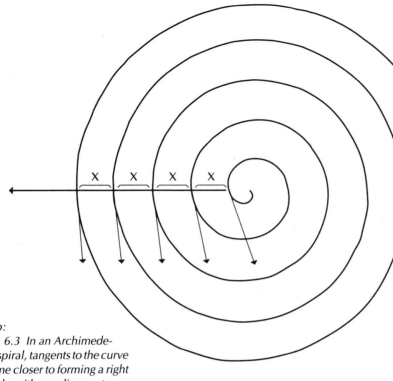

A mathematical peculiarity can be observed if a radius vector is drawn in an Archimedean spiral. The spiral intersects the radius vector at equally spaced intervals. Yet the angle formed by the spiral and the radius vector slowly changes, gradually approaching a right angle as the distance from the pole increases. The distances from the pole of the successive whorls of the spiral along a radius vector are in **arithmetic sequence**: x, 2x, 3x, 4x, ...

In Nature, some spiders use this spiral to spin their webs. They first lay out a framework of spokes which correspond to the radius vectors. They then move outward from the center to construct a spiral by guiding themselves by the previous turn so that the distance between the turns is everywhere the same.

Top:
Fig. 6.3 In an Archimedean spiral, tangents to the curve come closer to forming a right angle with a radius vector as the distance from the pole increases.

Bottom:
Spiral form exhibited in a spider web.

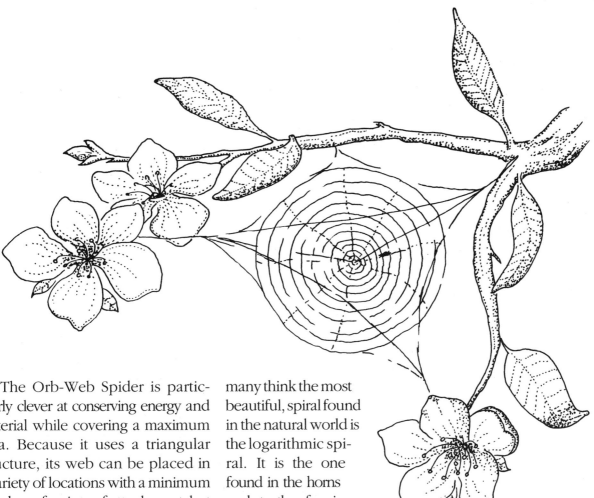

The Orb-Web Spider is particularly clever at conserving energy and material while covering a maximum area. Because it uses a triangular structure, its web can be placed in a variety of locations with a minimum number of points of attachment, but with maximum strength and flexibility. To build its web, the spider first lays down a bridge line between two vertical supports. It then attaches a strand near the middle of the bridge strand and reels out silk as it drops to another anchoring point, pulling on the original strand so that a Y shape is formed. The center of the Y becomes the center of the web so that there are three principal radii upon which the spider builds a spiral outwards.

Archimedean spirals also occur in the plant kingdom. They are seen in the whorl of young tomato plants, in the fiddlehead fern, and in the sago palm.

The Logarithmic Spiral

By far the most prevalent, and many think the most beautiful, spiral found in the natural world is the logarithmic spiral. It is the one found in the horns and teeth of animals; some spider webs; the seed cases and stems of plants; the beaks of birds; the shells of sea creatures; the umbilical cord; the cochlea of the ear; the great bones in humans; and the galaxies of the Universe.

This is a spiral of growth that exemplifies Nature's laws of conservation. It fills space economically and regularly, affords strength with a minimum of materials, but has the exceptional property that while expanding it alters its size but never its shape. Because of the aforementioned properties, the spiral provides a perfect solution to the problems of growth from a single point, growth by accretion, and growth under resistance, all factors which divert growth from a direct path.

Sea shells present the clearest examples of logarithmic spirals, but it is the shell of the chambered nautilus that comes closest to mathematical perfection. This 500 million year old creature can withstand pressure of one half ton per square inch because of its chambered construction. It begins life with a single chamber half as large as a pea and continues to add successively larger chambers every lunar month, accumulating as many as thirty in a lifetime. The animal lives only in the outer chamber. The preceding ones are first filled with gas and then sealed off to provide buoyancy which allows for locomotion through its environment at depths of several thousand feet. The animal's shell maintains a spiral shape with each new chamber being the gnomon for the entire form. It grows by accretion wherein secreted materials harden and can no longer change. Because these materials accumulate faster on the outside edge than on the inside, the shell curls in upon itself.

The shell of the adult nautilus can be likened to a cone that has been rolled up wherein the earliest and oldest chambers are the smallest. A great many other shells are essentially rolled tubes with the differences between species resulting from the differing rates of expansion of the coiled tubes.

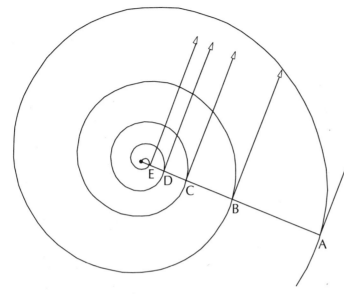

If we examine a cross section of this shell, we can see that it differs from the Archimedean Spiral in two ways. First, the spiral maintains a constant angle of intersection with the radius vectors, which prompted Descartes, in the 17th century, to name it the **equiangular spiral.**

In the 18th century, it was named the **proportional spiral** by Halley because of its other special property. That is, every logarithmic spiral is characterized by a particular geometric sequence, and the distances from the pole of the successive whorls along a radius vector correspond to the terms of that sequence. Thus the intervals increase proportionally further from the pole.

The 19th century mathematician Bernouilli called it both the *Spira Mirabilis* (miraculous spiral) and the logarithmic spiral. It is by the latter name that it is commonly known today.

Every logarithmic spiral is related to a characteristic rectangle. Within each member of the family of Dynamic Rectangles can be drawn a straight line or angular spiral using the diagonals of the rectangle and its reciprocal as the armature. Segments parallel to the sides are drawn between the intersecting diagonals. The nautilus shell described previously fits into the $\sqrt{\emptyset}$ Rectangle.

Fig. 6.5 illustrates this concept with some of the Dynamic Rectangles and extends the idea to encompass Dynamic Parallelograms and straight line spiral designs.

This page:

Fig. 6.4 In the logarithmic spiral $\frac{AB}{BC} = \frac{BC}{CD} = \frac{CD}{DE}$ and $\angle A \cong \angle B \cong \angle C \cong \angle D \cong \angle E$. The chambered nautilus fits within the $\sqrt{\emptyset}$ Rectangle.

This page:
Spiral forms on shells.

This page:
Drawing based on a meditation page from a 7th Century Celtic manuscript, Book of Durrow. In the original color work, the image is developed from one continuous curve.

Angular Spirals

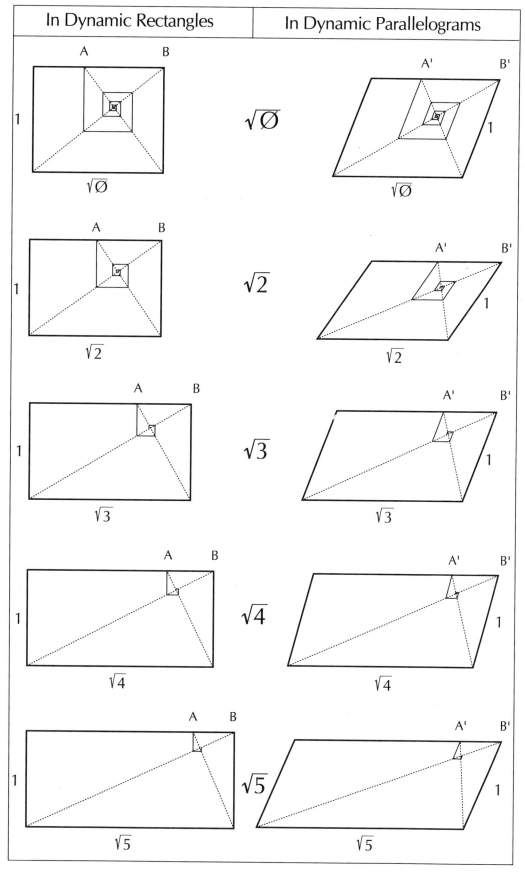

In Dynamic Rectangles	In Dynamic Parallelograms

This page:
Fig. 6.5   The Dynamic Parallelograms may be "tipped" at any angle. Their reciprocals are gotten by using the same points of intersection as in the parent rectangle.
eg. AB = A'B'

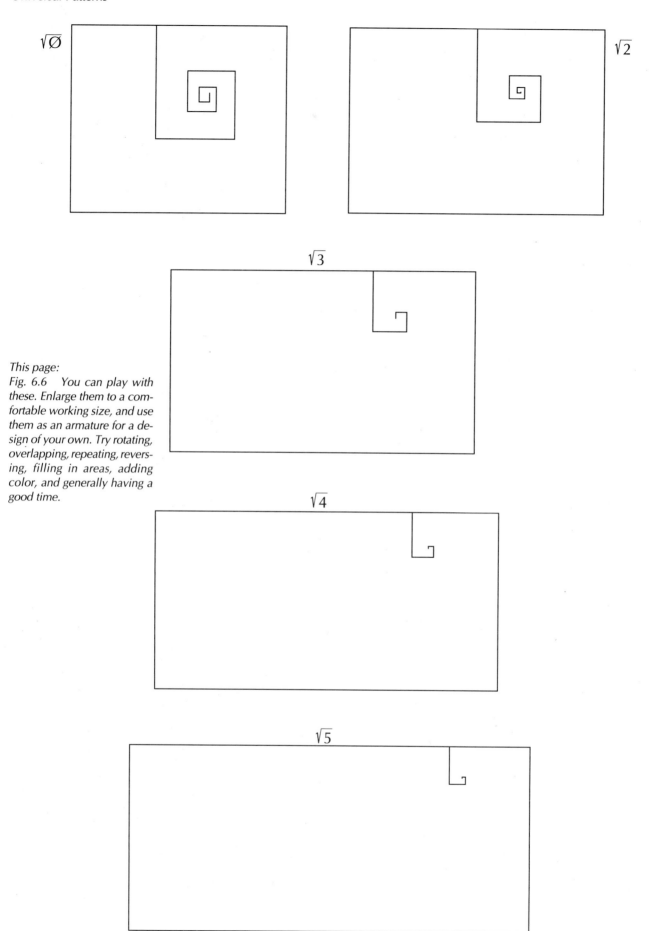

√Ø √2 √3

This page:
Fig. 6.6 You can play with these. Enlarge them to a comfortable working size, and use them as an armature for a design of your own. Try rotating, overlapping, repeating, reversing, filling in areas, adding color, and generally having a good time.

√4 √5

The Greek Technique

Notice in the previous diagram that the greater the ratio of length to width, the fewer whorls in the rectangle. That is, as the length increases in relation to a fixed width, the spiral moves away from the pole much more rapidly and therefore is less visible within the rectangle. The Classical Greeks were aware of this fact when creating spiral forms for the capitals of their temple columns. They experimented with different sized rectangles but found that the one that worked best for them was a 1.191 rectangle, which is one whose length lies between that of a square (1.000) and a $\sqrt{\varnothing}$ Rectangle (1.272). It seemed to allow for the correct number of whorls in relation to the column it capped.

Constructing a logarithmic spiral given just the pole and radius vectors is not so simple as it is with the Archimedean one. In his book, *Patterns and Design with Dynamic Symmetry*, Edward B. Edwards describes in detail the method for constructing logarithmic spirals within the Dynamic Rectangles using French curves and templates. We refer the reader to this resource for more detailed information.

The method used by the Greeks is by far less taxing mathematically, and offers a product quite sufficient for most design purposes. We shall examine the steps involved here. This method may be used with any rectangle, but we will demonstrate the procedure with the 1.191 rectangle which was the first choice of the Greeks.

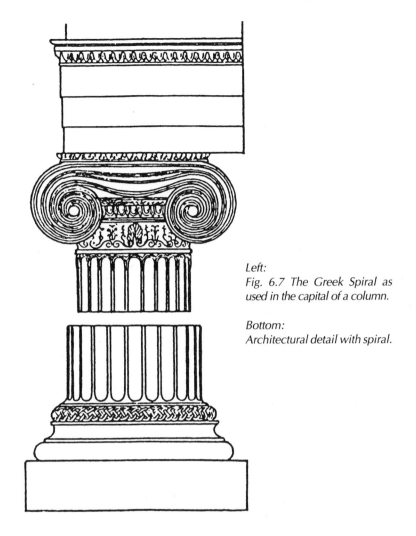

Left:
Fig. 6.7 The Greek Spiral as used in the capital of a column.

Bottom:
Architectural detail with spiral.

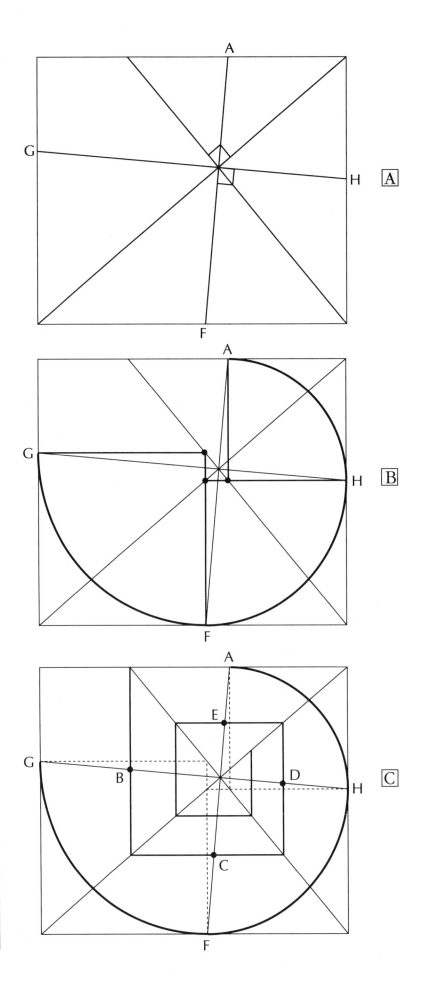

pp. 212-214:
Fig. 6.8 The Greek technique for creating spirals.

Top to bottom:
A. The main diagonal and the diagonal of the reciprocal (which is perpendicular to it) are drawn. The angles formed by their intersection are bisected resulting in $\overline{AF}$ and $\overline{GH}$. Note that the angle bisectors are not parallel to the sides.

B. Segments are drawn parallel to the sides of the rectangle from G, F, H, and A, respectively, to the diagonals as shown.
Three squares are formed which provide the framework for the first part of the spiral. Arcs are drawn through each square using the innermost vertex of each as the point upon which the tip of the compass is placed.

*C. The straight line, or angular, spiral is drawn as shown in Fig. 6.5. The spiral curve will be **tangent** to the sides of the angular spiral at points B, C, D and E, where it intersects the angle bisectors drawn in part A.*

Note: it would help to do this construction using layers of tracing paper so that earlier steps can be removed as you proceed. This reduces the visual clutter.

D

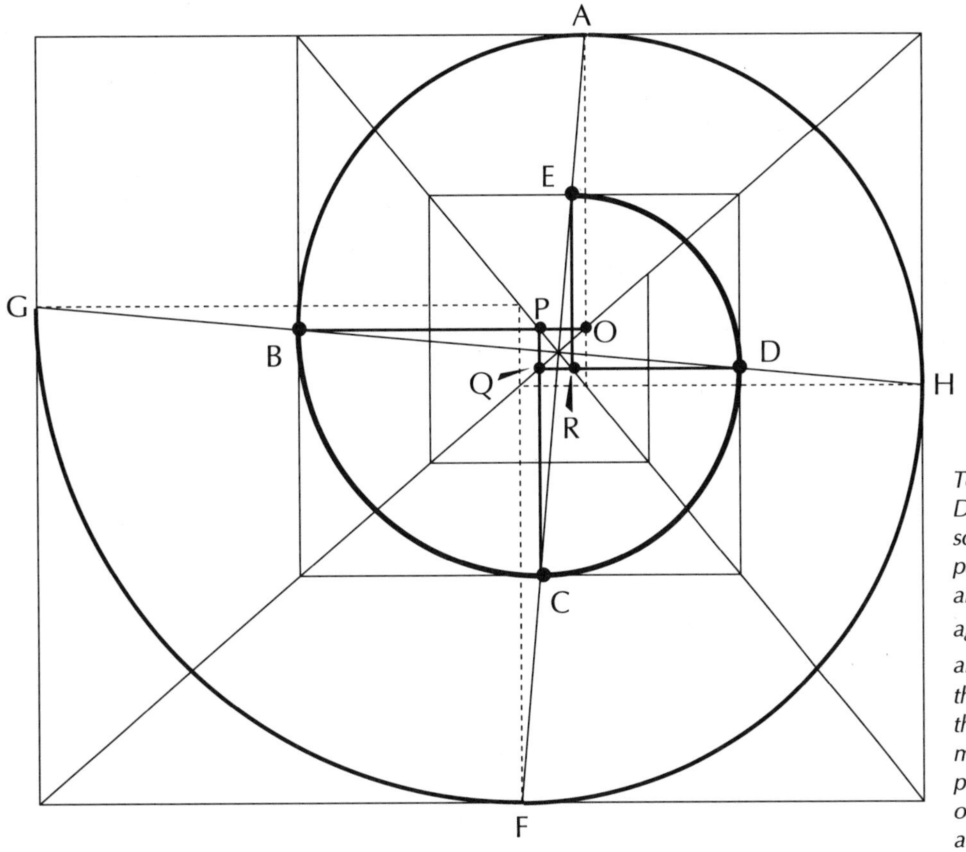

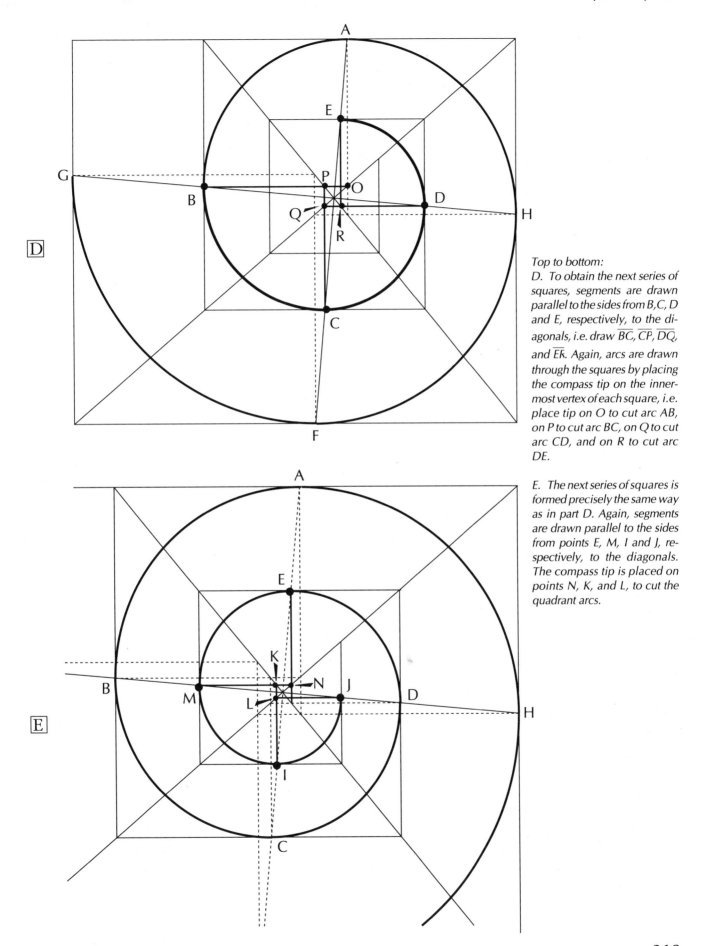

Top to bottom:
D. To obtain the next series of squares, segments are drawn parallel to the sides from B,C, D and E, respectively, to the diagonals, i.e. draw $\overline{BC}$, $\overline{CP}$, $\overline{DQ}$, and $\overline{ER}$. Again, arcs are drawn through the squares by placing the compass tip on the innermost vertex of each square, i.e. place tip on O to cut arc AB, on P to cut arc BC, on Q to cut arc CD, and on R to cut arc DE.

E. The next series of squares is formed precisely the same way as in part D. Again, segments are drawn parallel to the sides from points E, M, I and J, respectively, to the diagonals. The compass tip is placed on points N, K, and L, to cut the quadrant arcs.

E

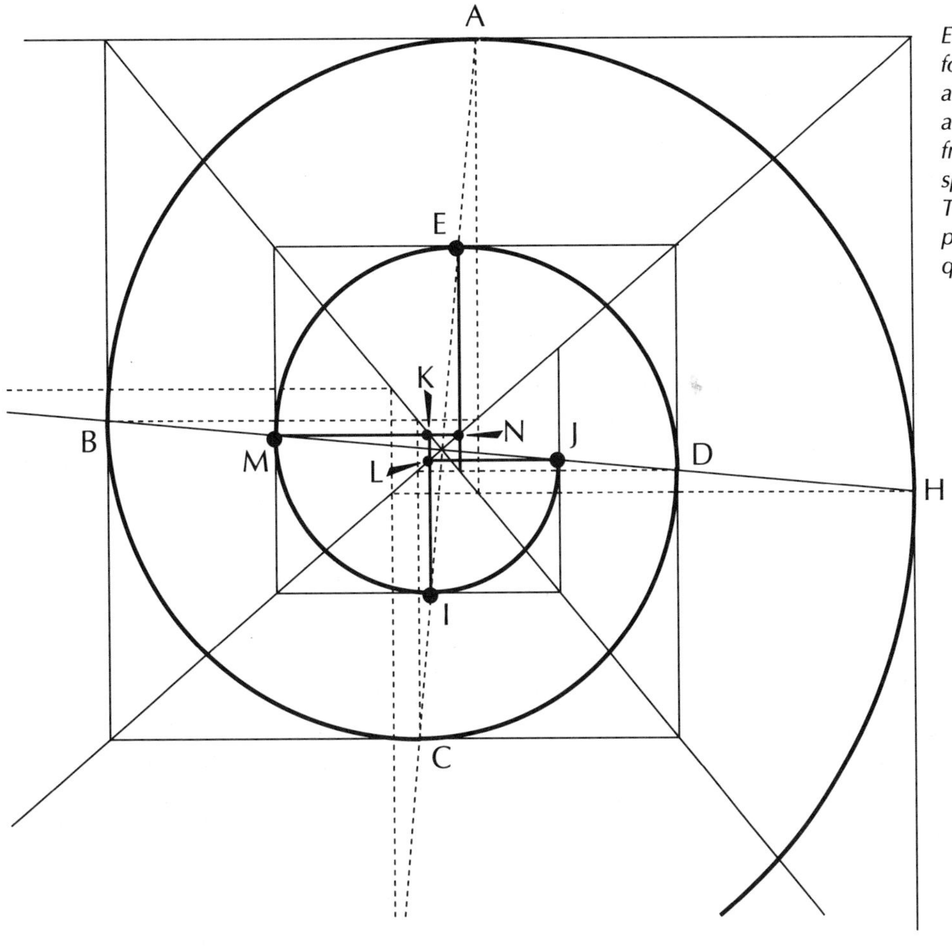

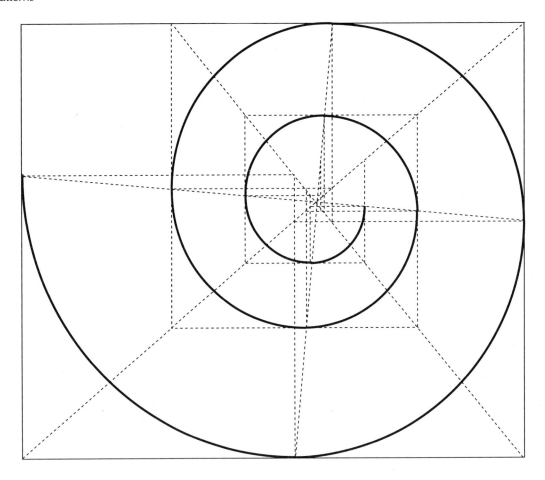

F

Top:
Fig. 6.8F The completed Greek spiral.

Bottom:
Concho spiral in ram's horns.

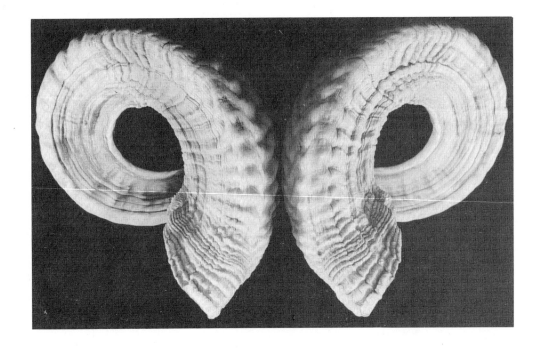

These spirals used by the Greeks were but approximations of true logarithmic spirals due to their method of construction. The one that felt "right" for them may have done so because it very closely imitates a logarithmic spiral with harmonic proportions. This one is called the Phidian spiral, again named for the Greek sculptor, because the lengths of the segments cut by successive whorls of the curve along any radius vector are in Ø proportion.

Because of its inherent pattern of "whirling squares", the Golden Rectangle provides the framework for a logarithmic-like spiral which is quite easy to obtain. This particular spiral is called the Golden Spiral and the method for its construction follows.

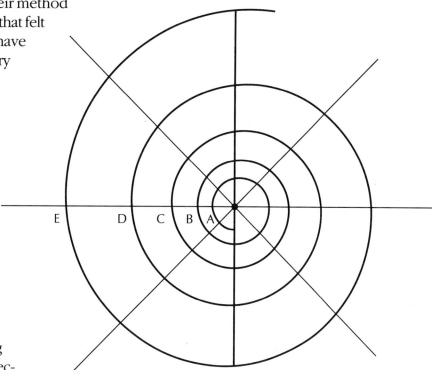

Above:
Fig. 6.9 The Phidian Spiral.
$$\frac{ED}{DC} = \frac{DC}{CB} = \frac{CB}{BA} = \emptyset \approx 1.61803$$

Below:
Fig. 6.10 The Golden Spiral in the Golden Rectangle.

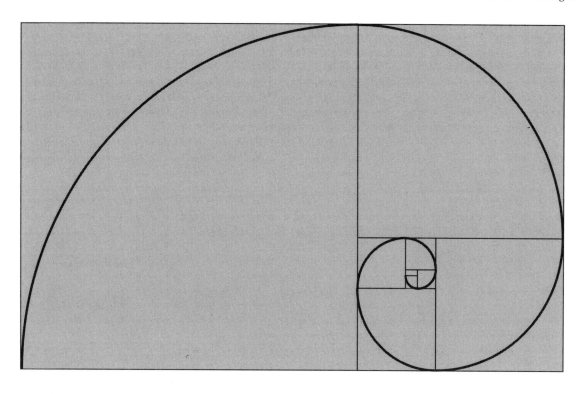

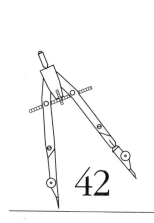

42

Construct a
Golden Spiral
Given a
Golden
Rectangle

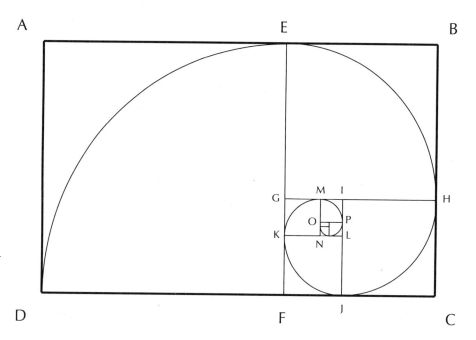

Given Golden Rectangle ABCD

1. Open the compass to measure AD. Place the metal tip on A and cut an arc on $\overline{AB}$.

2. Label the point of intersection E.

3. Without changing the setting, place the metal tip on D and cut an arc on $\overline{DC}$.

4. Label the point of intersection F.

5. Draw $\overline{EF}$.

Repeat Steps 1-5 in succession to get:

6. Square BHGE from rectangle BCFE.

7. Square CJIH from rectangle CFGH.

8. Square FKLJ from rectangle FGIJ.

9. Square GMNK from rectangle GILK.

10. Square MIPO from rectangle MILN.

This creates the pattern of *whirling squares.*

Note: AEFD is a square.

This page:
Wilcke Smith. Mixed media stitchery using ultrasuede fabric, bark paper, wool, needle-lace and gold wrapping.

11. Place the metal tip on O and draw arc PM.

12. Place the metal tip on N and draw arc MK.

13. Place the metal tip on L and draw arc KJ.

14. Place the metal tip on I and draw arc JH.

15. Place the metal tip on G and draw arc HE.

16. Place the metal tip on F and draw arc ED.

Now the curve is a Golden Spiral.

Note: the spiral may be enlarged by increasing the size of the Golden Rectangle using the method described on page 58, and continuing to cut the arc that joins the non-consecutive vertices of each new square.

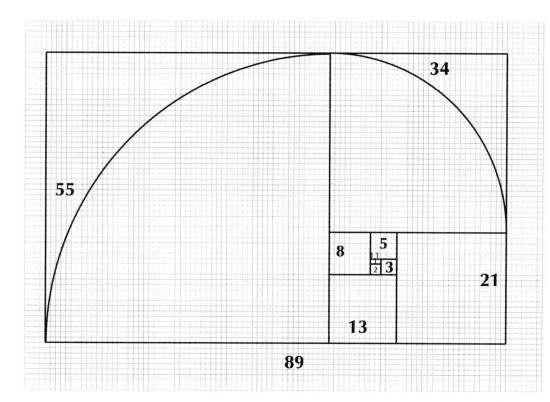

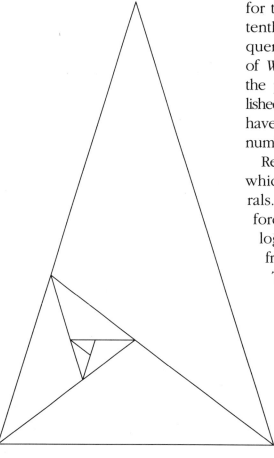

Above:
Fig. 6.11 The relationship between the Fibonacci numbers, the Golden Rectangle, and the Golden Spiral.

Right:
Fig. 6.12 The Golden Triangle.

The Golden Rectangle and the Golden Spiral bear an intimate relationship to those Golden Numbers, the Fibonacci Sequence. We know from Chapter 5 that consecutive terms of the Fibonacci Sequence yield a ratio very close to Ø. If we be gin with a rectangular grid and construct a rectangle whose length is 89 units and whose width is 55 units, we have a rectangle that closely approximates a Golden Rectangle. The length chosen for the sides are the eleventh and tenth terms of the Fibonacci Sequence (F_{11} and F_{10}). When the pattern of *Whirling Squares*, as described in the previous construction, is established, the successively smaller squares have sides named by the Fibonacci numbers in reverse sequence.

Rectangles are not the only figures which provide a framework for spirals. The Golden Triangle also affords the architecture for erecting a logarithmic spiral. You will recall from Chapter 4 that the Golden Triangle can be subdivided into a series of smaller and smaller Golden Triangles simply by bisecting successive base angles. Once this framework is established the related spiral, which approximates a logarithmic one, can be constructed.

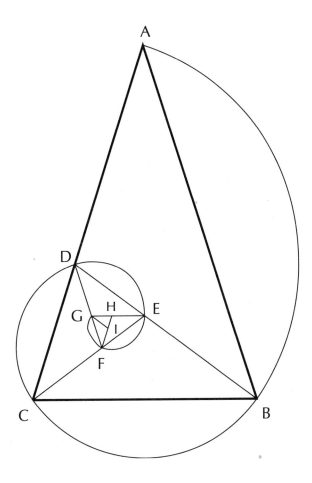

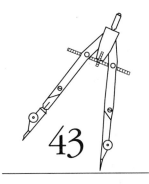

43

Construct a Spiral Given a Golden Triangle

Given Golden Triangle ABC

1. Subdivide the triangle by the method illustrated in Fig. 6.12 and label the points as shown.

2. Place the metal tip on D and draw arc AB.

3. Place the metal tip on E and draw arc BC.

4. Place the metal tip on F and draw arc CD.

5. Place the metal tip on G and draw arc DE.

6. Place the metal tip on H and draw arc EF.

7. Place the metal tip on I and draw arc FG.

Now the curve suggests a logarithmic spiral.

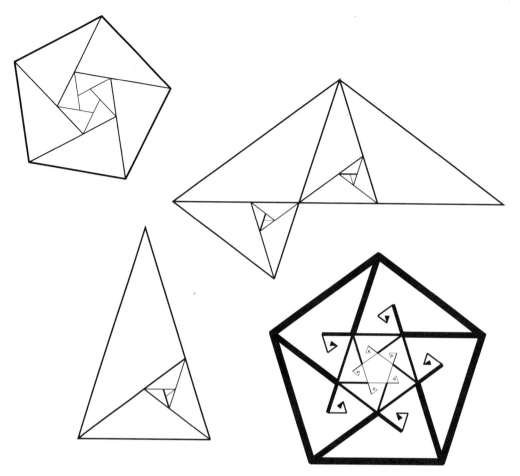

Top:
Fig. 6.13 The Golden Triangle may be combined with other figures to form particularly striking designs. In the examples given, harmony abounds in Ø proportions. Notice the use of two pentagons together with modified pentagrams which are formed by two rings of Golden Triangles. The straight line spirals follow the lines of the same subdivision of the Golden Triangle as in Construction 43.

Bottom:
Spiral-like detail from a piece of sculpture.

Spirals in the Third Dimension

Up to this point we have limited our discussion to spirals as they relate to the plane surface. As ours is a three-dimensional world, an analysis must be expanded to include the element of depth. Let us see what happens when the two spirals we have looked at are pulled up and out of the plane.

We shall first consider the case of the logarithmic spiral. In most natural examples the pole of the spiral is fixed in one plane and the curve expands not only away from the pole but through different planes as well. The resulting curve is called a **concho-spiral** and it is the one seen in many kinds of sea shells, in the scales of pine cones, and in some animal horns. When viewed head-on, the logarithmic spiral is clearly visible; when viewed from the side, the curve may pull sharply or gently away from the pole. The accompanying drawings illustrate these variations.

When the Archimedean spiral is pulled into the third dimension, a curve is formed for which the clearest example is the thread on a common wood screw. This curve may be described as the uniform wrapping of a line around a **cone** from its vertex to its base. The curve is called a **conical helix** and it is found less frequently in Nature than the conchospiral, but it is most useful in the world of human construction.

If the aforementioned line were to be wrapped around a **cylinder** rather than a cone, the result would be a cylindrical **helix**, or true helix. This curve, although spiral-like, is not truly a spiral since the diameter remains fixed as the curve lengthens. We include the helix in our discussion primarily because it is the curve fundamental to all life. The DNA molecule, which carries the genetic code for all living matter, is in the form of a double helix.

At a more visible level, we find the helix in the tendrils of hair and of vines. When expansion is needed in a confined space, the helix provides strength while expending the least amount of organic materials.

Near right:
Fig. 6.14 The conical helix.

Far right:
Fig. 6.15 The double helix, as found in the DNA molecule. This double spiral is considered a symbol of regeneration and a balancing of opposing movements.

In Buddhist philosophy, consciousness and its development is associated with the double spiral and its simultaneous inward and outward movement. For many religions, moving upward in a spiral path is the sense of the progression of the spirit turning from the material to the divine. One can spiral up to heaven or down to hell.

The life journey of each and every individual is a spiral path. Each person moves out from the center, which is the self, and moves round but never comes back to the same point.

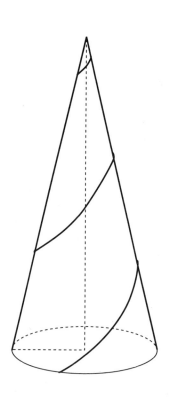

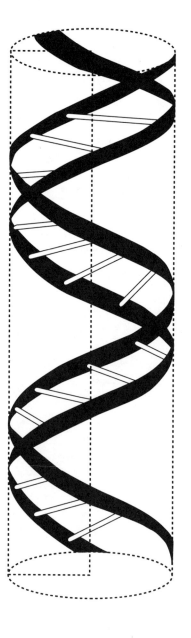

In human creation, the helix provides a practical as well as aesthetic solution to the problem of ascent in a space that is limited either horizontally or vertically. It gives us the spiral staircase, as found in French chateaux and English cathedrals, which can assume a gentle gradient and afford an easy climb.

Spirals in Art

The aesthetic sense recognizes and appreciates the fitness of certain forms and takes those as expressive of life, and, in the hands of the artist, gives them beauty. The artist searches out the essentials and refines the form to its elemental principles. Throughout the centuries the artists of the great civilizations have sought to understand the inherent qualities of natural forms and to express that understanding through their craft.

The spiral is one of the oldest recurring symbols crossing time and culture as far back as pre-history. As a symbol, it has the marvelous quality of holding many meanings simultaneously. It has been associated with the lotus blossom of Egypt, where a great many of the ornamental patterns were based on this natural form.

An abstract variation of the spiral, the swastika, is one of the most ancient and widespread symbols. It connotes both positive and negative forces. As a sun symbol the arms follow the direction of the sun as it moves through the four seasons. In this form, it is taken to be a token of good luck and is found on ancient pottery and artifacts from Mycenae and Troy in Greece. It has also been found among the ruins in the Yucatan in Mexico and on the woven blankets of the Navajo Indians of North America. It is also known as the Saxon fylfot and the Greek gammadion (a cross form of four capital gammas.) In the twentieth century, unfortunately, it was deliberately used in its negative form (with its arms twisting to the left) as a symbol of Nazi Germany.

Above:
Transformation from a more representational image of a lotus blossom to a stylized one in which the spiral form becomes evident.

Right:
Fig. 6.16
The swastika, an ancient symbol, is a variation on a spiral form.

In his book, *The Curves of Life*, Theodore Cook suggests that the aforementioned swastika is a derivation of a curved line, one which he feels evolved from the curves of the nautilus shell. The Oriental symbol of Yin-Yang, symbolizing the union of dark and light, male and female, positive and negative, may also be derived from this curve.

Simply drawing a "curly line" does not afford a believable facsimile of a spiral, and the artist-designer needs more concrete information at his or her disposal. The following pages demonstrate some of the ways that the spiral curve may be approximated without destroying its inherent characteristics.

Certain curves can be drawn which have the property of expanding in a regular fashion, but differ from a true spiral in that there is no single pole. Construction 44 demonstrates this type of curve generated by a square, hence having four "poles."

This page:
The spiral from the nautilus shell provides the basis for the curved swastika and yin-yang symbols.

225

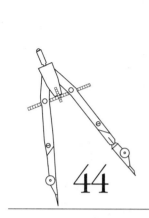

44

Construct a "Spiral" about a Square

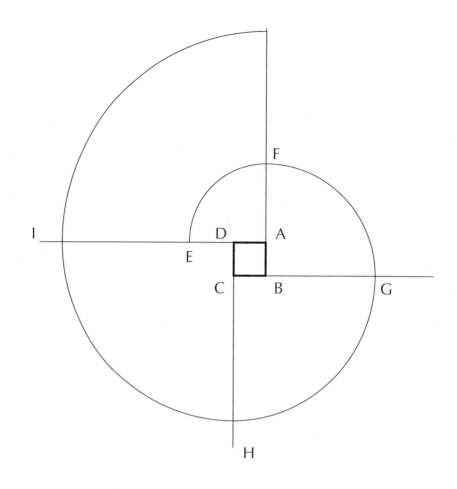

Given square ABCD

1. Extend $\overrightarrow{AD}$, $\overrightarrow{BA}$, $\overrightarrow{CB}$, and $\overrightarrow{DC}$.

2. Open the compass to measure more than AD, place the metal tip on A, and cut an arc that intersects $\overrightarrow{AD}$ and $\overrightarrow{BA}$. Label the points of intersection E and F, respectively.

3. Place the metal tip on B, the pencil tip on F, and cut an arc that intersects $\overrightarrow{CB}$. Label the point of intersection G.

4. Place the metal tip on C, the pencil tip on G, and cut an arc that intersects $\overrightarrow{DC}$. Label the point of intersection H.

5. Place the metal tip on D, the pencil tip on H, and cut an arc that intersects $\overrightarrow{AD}$. Label the point of intersection I.

6. Continue the process until the curve is the desired size.

Now the curve is spiral-like with the vertices of the square serving as four "poles".

Another way to obtain a curve similar to the ones produced by the method given in Construction 44 is to use two points on a line and alternate them as centers of semicircles as illustrated in the following construction.

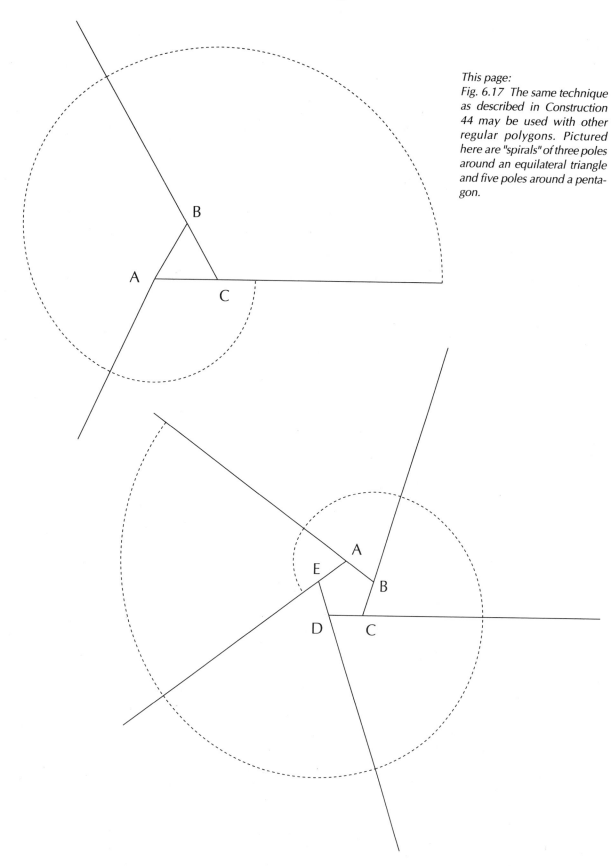

This page:
Fig. 6.17 The same technique as described in Construction 44 may be used with other regular polygons. Pictured here are "spirals" of three poles around an equilateral triangle and five poles around a pentagon.

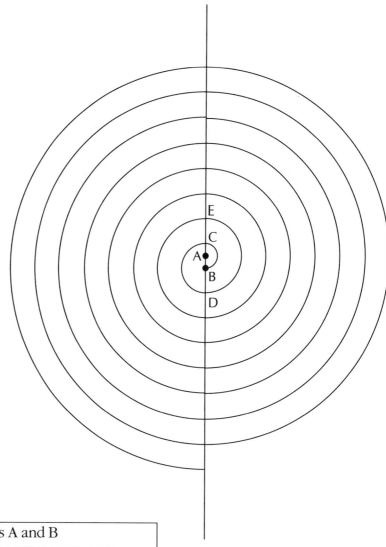

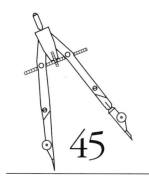

45

Construct a "Spiral" Using Alternate "Poles"

Given points A and B

1. Draw $\overleftrightarrow{AB}$.

2. Place the metal tip on A, the pencil tip on B, and cut semicircle BC.

3. Place the metal tip on B, the pencil tip on C, and cut semicircle CD.

4. Place the metal tip on A, the pencil tip on D, and cut semicircle DE.

5. Continue to cut semicircles on either side of $\overleftrightarrow{AB}$ by alternately plac-ing the metal tip on A and B until the curve is the desired size.

Now the curve is spiral-like with two "poles".

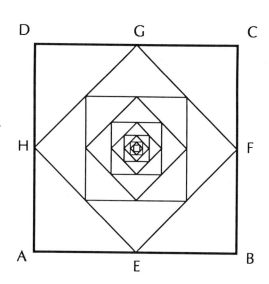

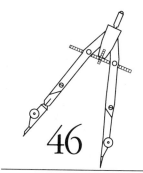

Construct a Baravelle Spiral within a Square

One of the simplest of spiral forms to construct is the Baravelle Spiral. It is a straight line design based on connecting consecutive mid-points of sides of regular polygons. Construction 46 gives the method for developing one within a square. The same concept may be applied to other regular polygons. Two examples are given in Fig. 6.18 on the following page.

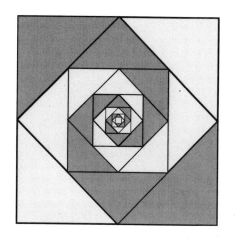

Given square ABCD

---

1. Locate the midpoints of $\overline{AB}$, $\overline{BC}$, $\overline{CD}$ and $\overline{DA}$.

2. Label them E, F, G and H, respectively.

3. Join them with line segments to form square EFGH.

4. In the same way, join the midpoints of the sides of EFGH to form another square.

5. Repeat the process until the final square is the desired size.

6. The "spiral" shape becomes visible when the triangles are shaded as illustrated in the diagram.

Now the resulting figure is a Baravelle Spiral.

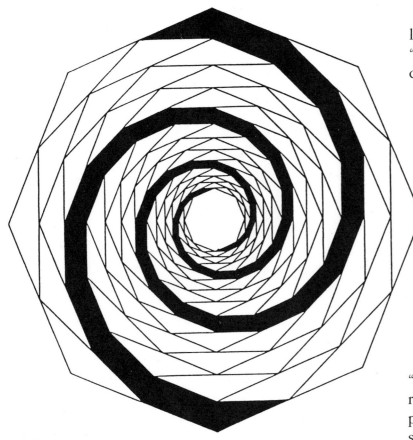

Baravelle Spirals have a very angular quality and, without shading, the "spiral" is not clearly evident. By choosing a point other than the midpoint of the sides, more subtle variations can be created which more nearly approximate a true curve. The closer the point is to an endpoint, the more curved the finished "spiral" will appear. In Construction 47, the point is located one-sixth of the distance from one vertex to the next. A greater or lesser distance may be chosen depending on whether a more angular look or a smoother appearance is desired.

Another advantage of this type of "spiral" is that the designer is not restricted to regular polygons. The procedure is demonstrated in the square in Construction 47. Other examples of straight line "spirals" are given in Fig. 6.19.

This page:
Fig. 6.18 Baravelle Spirals in other polygonal shapes.

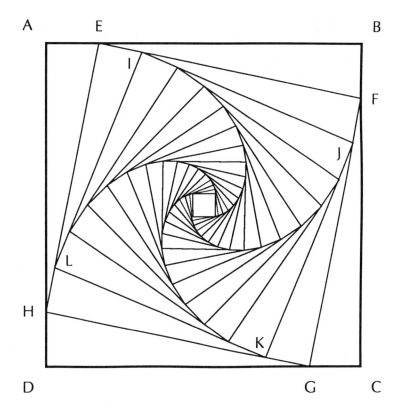

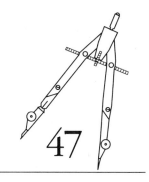

47

Construct a
Straight line
Spiral within a
Square

In this construction, four right triangles are formed at each step. When the sides of those triangles are triples of the form c −b = 1, the spiral is called a Pythagorean spiral.

Given square ABCD

1. Locate points E, F, G and H one-sixth (an arbitrary choice) of the distance from A to B, B to C, C to D, and D to A, r espectively. *(Use Construction 9 to divide a line segment into sixths.)* Connect the points to form square EFGH.

2. Find one-sixth the distance from E to F and use the measurement to mark off points I, J, K and L.

3. Connect the points to form square IJKL.

4. Repeat the process until the last square constructed is the desired size.

Now the figure is a straight line spiral.

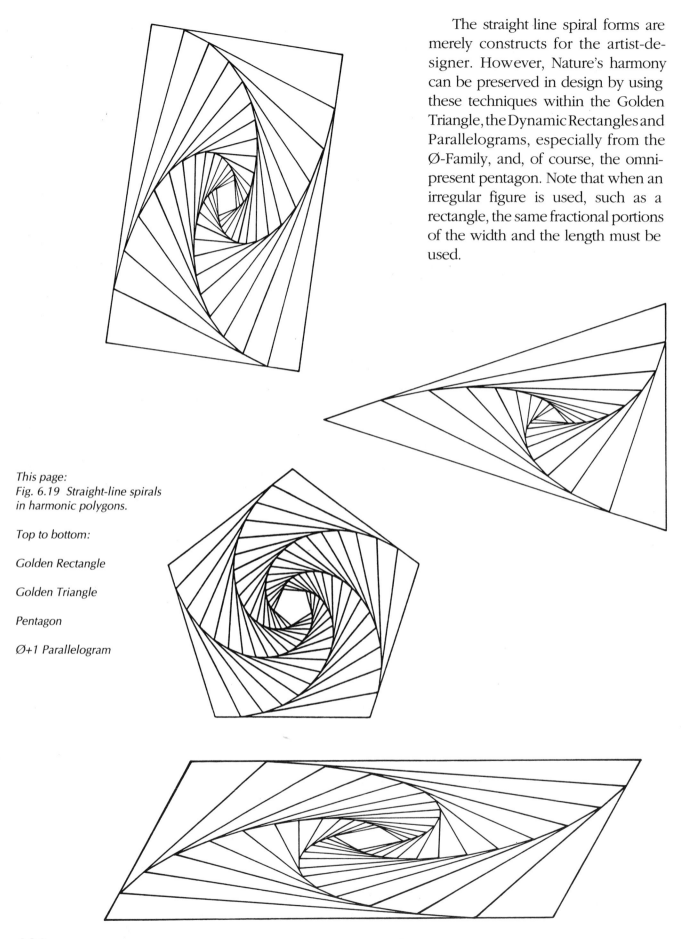

The straight line spiral forms are merely constructs for the artist-designer. However, Nature's harmony can be preserved in design by using these techniques within the Golden Triangle, the Dynamic Rectangles and Parallelograms, especially from the Ø-Family, and, of course, the omnipresent pentagon. Note that when an irregular figure is used, such as a rectangle, the same fractional portions of the width and the length must be used.

This page:
Fig. 6.19 Straight-line spirals in harmonic polygons.

Top to bottom:

Golden Rectangle

Golden Triangle

Pentagon

Ø+1 Parallelogram

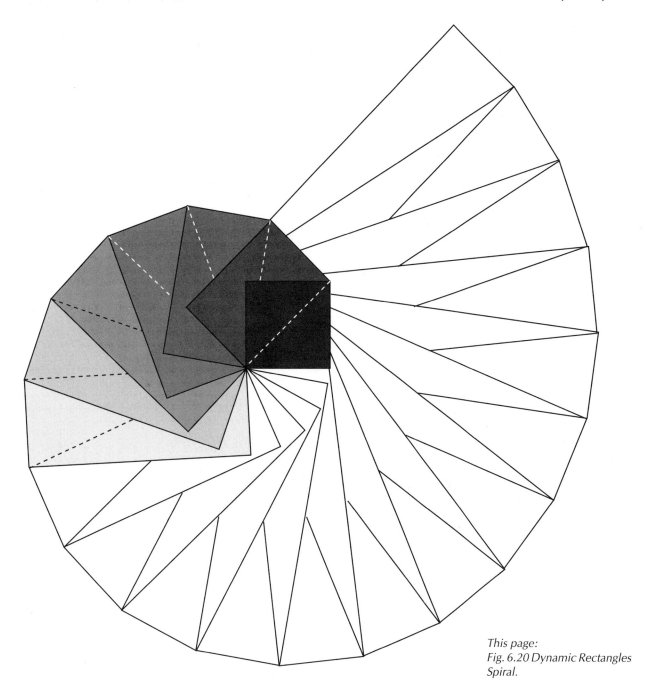

A last interesting straightline spiral can be constructed using a square and the resulting progression of Dynamic Rectangles. You will recall from Construction 31 that the length of each of the Dynamic Rectangles depends on the length of the previous rectangle. In this case, the vertex of the original unit square becomes the pole about which the Dynamic Rectangles rotate. Each new rectangle is constructed directly on the diagonal of the preceding one, with its width the same as the length of the side of the square.

This spiral is an example of a case where going beyond the $\sqrt{5}$ Rectangle is essential. Even if the process were carried on indefinitely, all of the Dynamic Rectangles could not be generated, but the spiral would eventually approach a smooth curve.

Problems

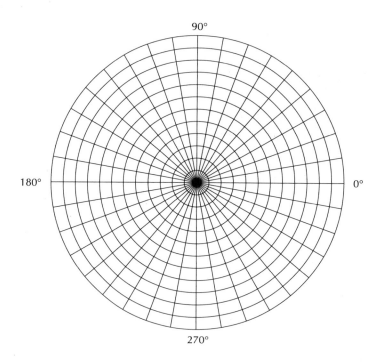

90°

180°   0°

270°

This page:
Fig. 6.21 Polar Graph. Use with Problem 1.

1 Use commercial polar graph paper or photocopy the one given here. Construct Archimedean spirals that:

a. increase one unit for each turn of 40°,

b. increase one half a unit for each turn of 60°, and

c. increase one unit for each turn of 90°.

2 Construct an angular spiral in any *two* of the Ø-Family rectangles.

3 Construct a Golden Spiral within a Golden Rectangle.

4 Construct two different sized Golden Triangles with their respective spirals.

5 Use the technique of Construction 44 to develop a "spiral" around a regular hexagon.

6 Construct a Baravelle Spiral in a decagon. *(You may discover a short cut in the process of doing this.)*

7 Locate two points one half an inch apart and use them as alternating "poles" to construct a "spiral."

8 Construct a Dynamic Rectangles Spiral beginning with a square no greater than one inch on a side.

9 Construct a spiral using the Greek Technique.

Projects

This page:
Fig. 6.22 Use with Project 1.

1 Using Fig. 6.22 as your inspiration, create an allover pattern with the Baravelle Spiral. Or develop a variation of your own in another regular polygon. Use a personal color structure for your finished work.

2 Construct a Baravelle or straight line spiral in any regular polygon and enhance it with a color scheme.

3 Use the concept of the spiral to create a three-dimensional mobile form.

4 Using the concept of the double helix, create a sculptural form.

5 Use the concept of the Dynamic Rectangles Spiral and a monochromatic, complementary, or triadic color scheme to develop a poster, a flag, or a banner.

235

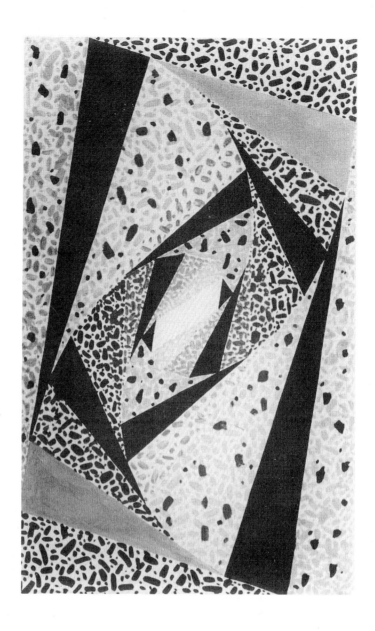

This page:
Lisa Petrucci. Student Work.
Chapter 6 Variation on Project
2.

6 Using what you know about the property of similarity, create a group of graduated pyramids based on the Golden Triangle. Join them together to make a three-dimensional spiral form. Color as you choose.

7 Create a composition using the concept of multi-poled "spirals".

8 Create a fantasy sea shell in two or three dimensions using a geometric progression in its whorls.

9 Do a drawing, painting, or collage that incorporates the following elements: flowers, butterflies, fish, and sea shells. Be concerned with their connections to harmonic forms.

10 Create a composition using Golden Spirals as they relate to either Golden Rectangles or Golden Triangles.

Further Reading

Bain, George. *Celtic Art, the Methods of Construction.* New York: Dover Publications, Inc., 1973.

Ball, Rouse. *A Short Account of the History of Mathematics.* New York: Dover Publications, Inc., 1960.

Cook, Theodore A. *The Curves of Life.* New York: Dover Publications, Inc., 1960.

Edwards, Edward B. *Pattern and Design with Dynamic Symmetry.* New York: Dover Publications, Inc., 1967.

Hargittai, Istvan and Clifford A. Pickover, Ed. *Spiral Symmetry.* River Edge, NJ: World Scientific Publishing Co. Pte. Ltd., 1992.

Huntley, H.E. *The Divine Proportion.* New York: Dover Publications, Inc., 1970.

Purce, Jill. *The Mystic Spiral.* New York: Thames and Hudson, 1974.

Stevens, Peter. *Patterns in Nature.* Boston: Little, Brown and Co., 1974.

Conclusion

In this book we have seen the Golden Ratio, Ø, to be the common thread that weaves harmony into the universal patterns. This section is presented to visually clarify the many connections.

The Golden Ratio

$$Ø ≈ 1.61803$$

The Divine Proportion

$$\frac{AB}{AC} = \frac{AC}{CB} = Ø$$

or

$$\frac{a+b}{a} = \frac{a}{b} = Ø$$

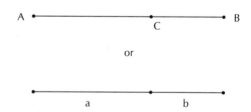

The Fibonacci Sequence: The Golden Numbers

Beyond F_{15} the ratio of any pair of consecutive Fibonacci numbers is the same as Ø to five decimal places.

n	F_n	n	F_n
1	1	9	34
2	1	10	55
3	2	11	89
4	3	12	144
5	5	13	233
6	8	14	377
7	13	15	610
8	21	16	987

The Golden Sequence

$$1, \emptyset, \emptyset^2, \emptyset^3, \emptyset^4, \emptyset^5, \ldots$$

This sequence can be rewritten in the following way to show its relationship to the Fibonacci sequence:

$$1, \emptyset, 1+\emptyset, 1+2\emptyset, 2+3\emptyset, 3+5\emptyset, 5+8\emptyset, \ldots$$

The Golden Rectangle

$$\frac{AB}{BC} = \frac{BC}{FB} = \frac{FB}{BH} = \emptyset$$

This is the *only* rectangle whose gnomon is a square, hence the name whirling square rectangle.

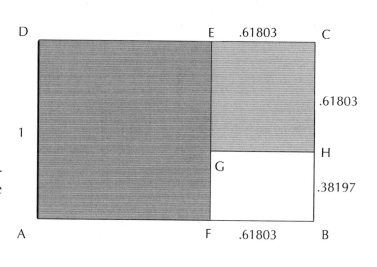

The Golden Rectangle in relationship to the Fibonacci numbers

The Golden Rectangle together with the Golden Spiral

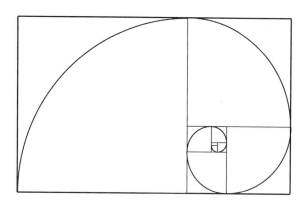

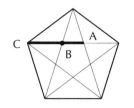

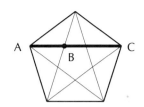

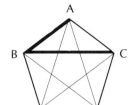

The Golden Ratio in the Golden Polygon: the Pentagon with Pentagram

In each case

$$\frac{AB+BC}{BC} = \frac{BC}{AB} = \emptyset$$

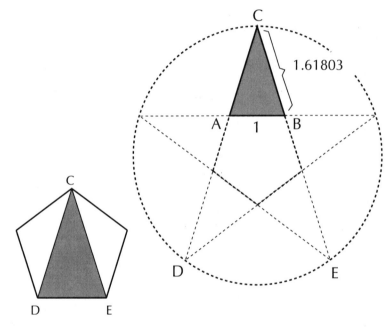

The Golden Triangle

In the Pentagon and Pentagram

$$\frac{CB}{AB} = \frac{CE}{DE} = \emptyset$$

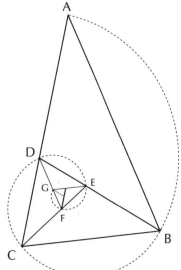

With its Logarithmic Spiral

$$\frac{AB}{BC} = \frac{BC}{CD} = \frac{CD}{DE} = \frac{DE}{EF} = \frac{EF}{FG} = \emptyset$$

In the Lute of Pythagoras

$$\frac{AB}{BC} = \frac{BC}{CD} = \frac{CD}{DE} = \frac{DE}{EF} = \varnothing$$

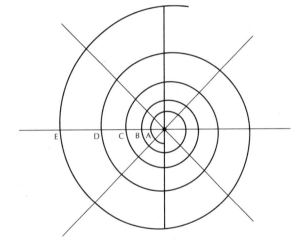

The Phidian Spiral

$$\frac{ED}{DC} = \frac{DC}{CB} = \frac{CB}{BA} = \varnothing$$

The Phi-Family Rectangles

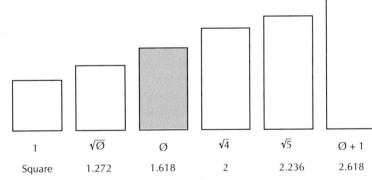

1	$\sqrt{\varnothing}$	$\varnothing$	$\sqrt{4}$	$\sqrt{5}$	$\varnothing + 1$
Square	1.272	1.618	2	2.236	2.618

We have come to the end of Book I. We hope that the reader has gained insight and, like us, has been filled with a deeper appreciation of the omnipresent connections among objects and events of this universe.

Selected References

Ball, Rouse. *A Short Account of the History of Mathematics*. New York: Dover, 1960.

Beard, Co. Robert S. *Patterns in Space*. Palo Alto: Creative Publications, 1973.

Bentley, W.A. and W.J. Humphreys. *Snow Crystals*. New York: Dover, 1962.

Bezuska, Stanley, Margaret Kenney, Linda Silvey. *Designs From Mathematical Patterns*. Palo Alto: Creative Publications, 1978.

Bouleau, Charles. *The Painter's Secret Geometry*. New York: Harcourt, Brace and World, 1963.

Bruni, James. *Experiencing Geometry*. Belmont: Wadsworth, 1977.

Capra, Fritjof. *The Tao of Physics*. New York: Bantam Books, 1965.

Capra, Fritjof. *The Turning Point*. New York: Bantam Books, 1983.

Colman, Samuel. *Nature's Harmonic Unity*. New York: Putnam's and Sons, 1911.

Cook, Theodore A. *The Curves of Life*. New York: Dover, 1979.

Coxeter, H.S.M. *Introduction to Geometry*. New York: John Wiley and Son, 1961.

Critchlow, Keith. *Order in Space*. New York: The Viking Press, 1970.

Cundy and Rollett. *Mathematical Models*. New York: Oxford University Press, 1961.

Doczi, Gyorgy. *The Power of Limits*. Boulder: Shambala Press, 1981.

Edwards, Edward B. *Pattern and Design with Dynamic Symmetry*. New York: Dover, 1967.

Feininger, Andreas. *The Anatomy of Nature*. New York: Dover, 1956.

Fleming, William. *Arts and Ideas*. New York: Holt, Rinehart and Winston, 1980.

Garland, Trudi Hammel. *Fascinating Fibonaccis*. Palo Alto: Dale Seymour Publications, 1987.

Ghyka, Matila. *The Geometry of Art and Life*. New York: Dover, 1977.

Gillings, Richard J. *Mathematics in the Time of the Pharaohs*. New York: Dover, 1972.

Grillo, Paul Jacques. *Form, Function and Design*. New York, Dover, 1960.

Haeckel, Ernst. *Art Forms in Nature*. New York: Dover, 1974.

Hambidge, Jay. *The Elements of Dynamic Symmetry*. New York: Dover, 1967.

Hambidge, Jay. *The Parthenon and Other Greek Temples, Their Dynamic Symmetry*. New Haven: Yale University Press, 1924.

Hambidge, Jay. *Practical Applications of Dynamic Symmetry*. New Haven: Yale University Press, 1932.

Hogben, Lancelot. *Mathematics in the Making*. New York: Doubleday, 1960-61.

Hogben, Lancelot. *The Wonderful World of Mathematics.* New York: Garden City Books, 1955.

Holden, Alan. *Shapes, Space and Symmetry.* New York: Columbia University Press, 1971.

Huntley, H.E. *The Divine Proportion.* New York: Dover, 1970.

Kline, Morris. *Mathematics and the Physical World.* New York: Oxford University Press, 1953.

Kline, Morris. *Mathematics in Western Culture.* New York: Oxford University Press, 1953.

Pearce, Peter and Susan Pearce. *Polyhedra Primer.* New York: Van Nostrand Reinhold, 1978.

Runion, Garth E. *The Golden Section.* Palo Alto: Dale Seymour Publications,1990.

Seymour, Dale and Reuben Schadler. *Creative Constructions.* Palo Alto: Creative Publications, 1974.

Stevens, Peter S. *Patterns in Nature.* Boston: Little, Brown and Co., 1974.

Thompson, D'Arcy. *On Growth and Form.* Cambridge: University Press, 1969.

Wenninger, Magnus J. *Polyhedron Models.* Cambridge: University Press, 1971.

Weyl, Hermann. *Symmetry.* Princeton: University Press, 1952.

Williams, Christopher. *Origins of Form.* New York: Visual Communications Books, 1981.

A Mathematical Symbols

$\overline{AB}$ segment AB

$\overleftrightarrow{AB}$ line AB

$\overrightarrow{AB}$ ray AB

AB length of segment AB (designates a real number)

$\triangle ABC$ triangle ABC

$\angle ABC$ angle ABC

$\cong$ is congruent to

$=$ is equal to

$\approx$ is approximately equal to

$\sim$ is similar to

$\perp$ is perpendicular to

$>$ is greater than

$\geq$ is greater than or equal to

$<$ is less than

$\leq$ is less than or equal to

$\neq$ is not equal to

$\pm$ plus or minus (both operations designated by a single symbol)

$\therefore$ therefore

$x°$ x degrees

y' y minutes (angle measure)

z'' z seconds (angle measure)

x' x feet (linear measure)

y'' y inches (linear measure)

π pi ≈ 3.14 or 22/7 (the ratio of the circumference of a circle to its diameter)

$\varnothing$ phi ≈ 1.61803 (the Golden Ratio)

x_n x sub n (a way of naming a variable)

x^n x to the nth power (multiply x by itself n times)

B *Properties and Proofs*

Contents

Properties of Proportions

1 In any proportion, the product of the means equals the product of the extremes.

If $\dfrac{a}{b} = \dfrac{c}{d}$, then $ad = bc$.

2 If the product of one pair of numbers equals the product of another pair of numbers, either pair may be made the means and the other pair the extremes of a proportion.

If $ad = bc$, then $\dfrac{a}{b} = \dfrac{c}{d}$ or $\dfrac{b}{a} = \dfrac{d}{c}$.

3 In any proportion, the means or the extremes may be interchanged.

If $\dfrac{a}{b} = \dfrac{c}{d}$, then $\dfrac{a}{c} = \dfrac{b}{d}$ or $\dfrac{d}{b} = \dfrac{c}{a}$.

4 If four quantities are in proportion, they are in proportion by inversion or alternation.

If $\dfrac{a}{b} = \dfrac{c}{d}$, then $\dfrac{b}{a} = \dfrac{d}{c}$ or $\dfrac{c}{a} = \dfrac{d}{b}$.

Properties of Geometric Figures

Parallel Lines

1 If l is parallel to m and l and m are cut by transversal t, then:

$$\angle 1 \cong \angle 3 \cong \angle 5 \cong \angle 7$$

and

$$\angle 2 \cong \angle 4 \cong \angle 6 \cong \angle 8.$$

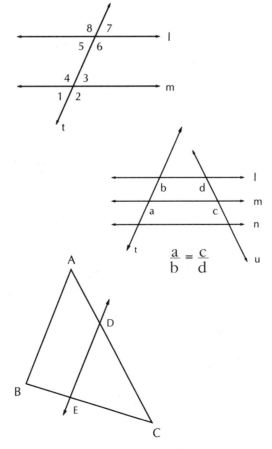

2 Three or more parallels cut proportional segments on two transversals.

If l, m, and n are parallel and cut by transversals t and u then the segments of the transversals are in proportion:

$$\frac{a}{b} = \frac{c}{d}$$

3 If a line is parallel to one side of a triangle, it divides the other two sides proportionally.

If $\overleftrightarrow{DE}$ and $\overleftrightarrow{AB}$ are parallel, then

$$\frac{AD}{BE} = \frac{DC}{EC}.$$

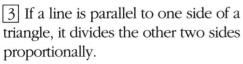

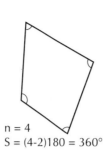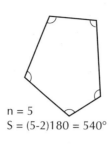

Polygons

$\boxed{1}$ The sum, S, of the interior angles of a polygon of n sides is S = (n-2)180.

n = 3
S = (3-2)180 = 180°

n = 4
S = (4-2)180 = 360°

n = 5
S = (5-2)180 = 540°

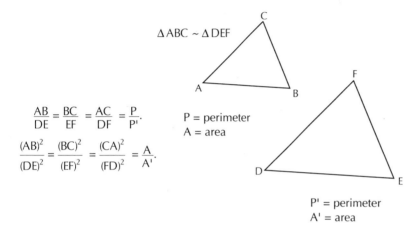

△ ABC ~ △ DEF

$$\frac{AB}{DE} = \frac{BC}{EF} = \frac{AC}{DF} = \frac{P}{P'}.$$

$$\frac{(AB)^2}{(DE)^2} = \frac{(BC)^2}{(EF)^2} = \frac{(CA)^2}{(FD)^2} = \frac{A}{A'}.$$

P = perimeter
A = area

P' = perimeter
A' = area

$\boxed{2}$ The perimeters of two similar polygons are proportional to the corresponding sides, but their areas are proportional to the squares of the corresponding sides.

Triangles

$\boxed{1}$ Two triangles are congruent if the following corresponding parts are congruent:

 a. one side and two corresponding angles, (ASA, SAA*),

 b. two sides and the included angle, (SAS),

 c. two sides and the right angle opposite one of them, (HL),

 d. three sides, (SSS).

*The abbreviations name the corresponding congruent parts: *S* means side, *A* means angle, *L* means leg, and *H* stands for hypotenuse. The parts are named in order as one travels around the triangle in either a clockwise or counterclockwise direction.

 In each case, the corresponding

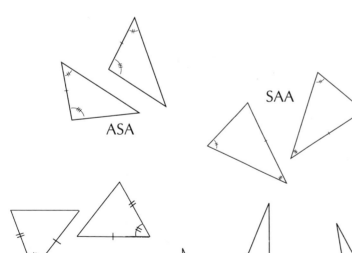

ASA

SAA

SAS

HL

SSS

congruent parts of the triangles are sufficient to ensure the congruence of the triangles themselves.

2 Two triangles are similar if:

a. two angles of one triangle are congruent to two angles of the other,

b. an angle of one triangle is congruent to an angle of the other and the corresponding including sides are proportional,

c. three sides of one triangle are in proportion to three sides of the other,

d. three sides of one triangle are parallel to three sides of the other. This is illustrated in example b.

In each case, the listed conditions are sufficient to ensure the similarity of the triangles.

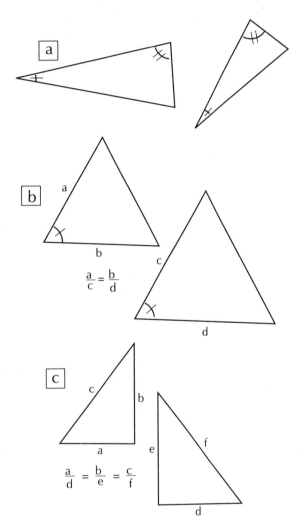

3 The altitudes of a triangle meet in a single point. See *altitude* in the glossary for a diagram.

4 The medians of a triangle meet in a point that is two-thirds of the distance from each vertex to the midpoint of the opposite side.

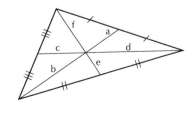

b=2a
d=2c
f=2e

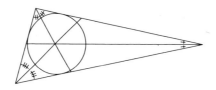

[5] The angle bisectors of a triangle meet in a point that is the center of the inscribed circle.

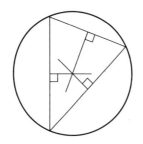

[6] The perpendicular bisectors of the sides of a triangle meet in a point that is the center of the circumscribed circle.

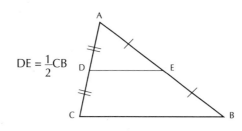

$DE = \frac{1}{2}CB$

[7] The line segment between the midpoints of two sides of a triangle is parallel to the third side and half as long.

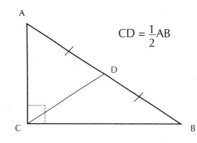

$CD = \frac{1}{2}AB$

[8] The median to the hypotenuse of a right triangle is half as long as the hypotenuse.

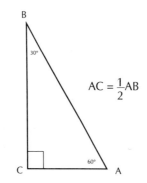

$AC = \frac{1}{2}AB$

[9] In a 30-60-90 Triangle, the side opposite the 30° angle is half the length of the hypotenuse.

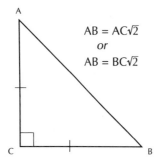

$AB = AC\sqrt{2}$
or
$AB = BC\sqrt{2}$

[10] In an isosceles Right Triangle, the length of the hypotenuse equals the length of a leg times $\sqrt{2}$.

Quadrilaterals

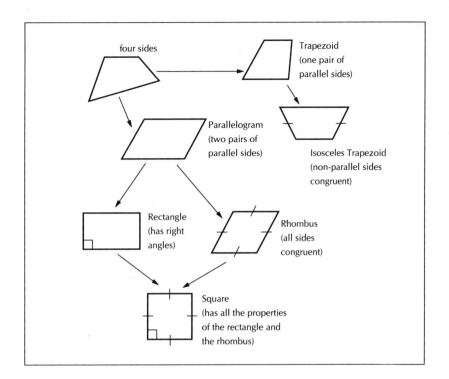

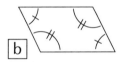

[1] In a parallelogram:
   a. opposite sides are congruent,
   b. opposite angles are congruent,
   c. the diagonals bisect each other, and
   d. a diagonal divides the parallelogram into two congruent triangles.

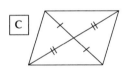

$\triangle ABD \cong \triangle CDB$

[2] The diagonals of a rectangle are congruent.

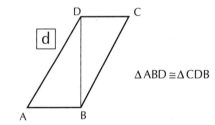

$\overline{AC} \cong \overline{DB}$

[3] The diagonals of a rhombus are perpendicular.

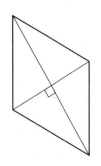

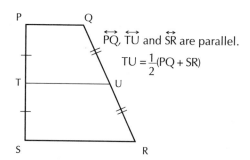

$\overleftrightarrow{PQ}$, $\overleftrightarrow{TU}$ and $\overleftrightarrow{SR}$ are parallel.

$TU = \frac{1}{2}(PQ + SR)$

[4] The line segment connecting the midpoints of the non-parallel sides of a trapezoid is parallel to the bases and half as long as the sum of their lengths.

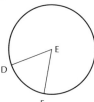

∠ SPQ ≅ ∠ PQR and ∠ PSR ≅ ∠ QRS.
$\overline{PR} \cong \overline{SQ}$.

[5] In an isosceles trapezoid:
   a. the base angles are congruent,
   b. the diagonals are congruent.

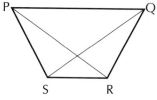

Assume circles B and E are congruent. If ∠ABC ≅ ∠DEF then arc AC ≅ arc DF , and if arc AC ≅ arc DF then ∠ABC ≅ ∠DEF .

Circles

[1] In the same circle or in congruent circles, congruent central angles cut congruent arcs, and conversely.

If $\overline{PQ} \cong \overline{RS}$ then arc PQ ≅ arc RS, and if arc PQ ≅ arc RS then $\overline{PQ} \cong \overline{RS}$.

[2] In the same circle or in congruent circles, congruent chords cut congruent arcs, and conversely.

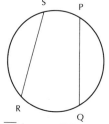

If $\overline{RS} \cong \overline{ST}$ then $\overline{XP} \cong \overline{PY}$.

[3] In the same circle or in congruent circles, congruent chords are equidistant from the center.

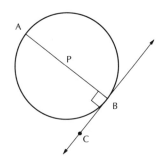

4 A line perpendicular to a diameter at one endpoint is tangent to the circle, and conversely.

If $\overleftrightarrow{AB} \perp \overleftrightarrow{BC}$ then $\overleftrightarrow{BC}$ is tangent to circle P, and if $\overleftrightarrow{BC}$ is a tangent line, then $\overleftrightarrow{AB} \perp \overleftrightarrow{BC}$.

5 Angles inscribed in the same arc or congruent arcs are congruent.

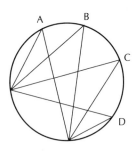

$\angle A \cong \angle B \cong \angle C \cong \angle D.$

6 Any angle inscribed in a semicircle is a right angle.

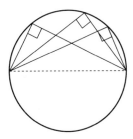

7 A circle, and only one circle, can be circumscribed about or inscribed in any triangle or any regular polygon.

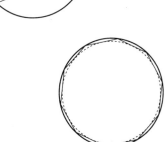

Properties of Exponents

1 If n is a positive integer,
$a^n = a \cdot a \cdot a \cdots a$ (n times)
eg. $3^4 = 3 \cdot 3 \cdot 3 \cdot 3 = 81$
Note: The exponent only operates on the quantity immediately to its left.
eg. $2x^3 = 2xxx$ but $(2x)^3 = 2x2x2x = 8x^3$.

2 $a^0 = 1$ for all real numbers $a \neq 0$.
0^0 is undefined.
eg. $1{,}532{,}896^0 = 1$.

3 $a^{-n} = \dfrac{1}{a^n}$ for all real numbers $a \neq 0$.
eg. $5^{-3} = \dfrac{1}{5^3} = \dfrac{1}{125}$.

4 $(a^n)(a^m) = a^{m+n}$.
eg. $\emptyset^2 \cdot \emptyset^5 = \emptyset^{2+5} = \emptyset^7$.

5 $\dfrac{a^n}{a^m} = a^{n-m}$ for all real numbers $a \neq 0$.
eg. $\dfrac{x^5}{x^3} = x^{5-3} = x^2$,
or $\dfrac{s^4}{s^5} = s^{4-5} = s^{-1} = \dfrac{1}{s}$.

6 $(a^m)^n = a^{mn}$.
eg. $(n^2)^3 = n^{2\times3} = n^6$.

7 $a^{p/q} = \left(\sqrt[q]{a}\right)^p = \sqrt[q]{a^p}$
eg. $8^{2/3} = \left(\sqrt[3]{8}\right)^2 = 2^2 = 4$.
$8^{2/3} = \sqrt[3]{8^2} = \sqrt[3]{64} = 4$.

Properties of Phi

1 $\emptyset - 1 = \dfrac{1}{\emptyset}$.

Proof: We know from the Divine Proportion that $\emptyset^2 - \emptyset - 1 = 0$, or $\emptyset^2 - \emptyset = 1$. Dividing both sides by $\emptyset$ we get
$\dfrac{\emptyset^2 - \emptyset}{\emptyset} = \dfrac{1}{\emptyset}$.

But also, $\dfrac{\emptyset^2 - \emptyset}{\emptyset} = \dfrac{\emptyset^2}{\emptyset} - \dfrac{\emptyset}{\emptyset} = \emptyset - 1$.
Therefore, $\emptyset - 1 = \dfrac{1}{\emptyset}$.

2 $\emptyset\emptyset' = -1$.

Proof:

$\emptyset\emptyset' = \left(\dfrac{1+\sqrt{5}}{2}\right)\left(\dfrac{1-\sqrt{5}}{2}\right) = \dfrac{1-5}{4} = \dfrac{-4}{4} = -1$.

3 $\emptyset + \emptyset' = 1$.

Proof:

$\emptyset + \emptyset' = \dfrac{1+\sqrt{5}}{2} + \dfrac{1-\sqrt{5}}{2} = \dfrac{2}{2} = 1$.

4 $\emptyset^n + \emptyset^{n+1} = \emptyset^{n+2}$ for n an integer.

Proof: First consider the case where $n \geq 0$. Then $\emptyset^n + \emptyset^{n+1} = \emptyset^{n+2}$ for $n \geq 0$ is equivalent to $\emptyset^{m-1} + \emptyset^m = \emptyset^{m+1}$ for $m \geq 1$.

(a) Using the method of mathematical induction, we first prove the statement true for m=1.

We know that $\emptyset^2 - \emptyset - 1 = 0$,

so $1 + \emptyset = \emptyset^2$, or $\emptyset^0 + \emptyset^1 = \emptyset^2$. Therefore true for $m = 1$.

(b) Assume true for $m = k$. Then $\emptyset^{k-1} + \emptyset^k = \emptyset^{k+1}$.

(c) Prove true for $m = k+1$. Since $\emptyset^{k-1} + \emptyset^k = \emptyset^{k+1}$, $\emptyset(\emptyset^{k-1} + \emptyset^k) = \emptyset\emptyset^{k+1}$, or $\emptyset^k + \emptyset^{k+1} = \emptyset^{k+2}$. Therefore, the statement is true for $m = k+1$, and, by the principle of mathematical induction, it holds for all positive integral values of m.

Now let us consider the case where $n < 0$.

If $n < 0$, then $\emptyset^n + \emptyset^{n+1} = \emptyset^{n+2}$ is equivalent to $\dfrac{1}{\emptyset^m} + \dfrac{1}{\emptyset^{m-1}} = \dfrac{1}{\emptyset^{m-2}}$ where $m = -n > 0$.

(a') Again using mathematical induction, we first prove the statement true for $m = 1$.

We know that $\emptyset$ differs from its reciprocal by 1, that is $\dfrac{1}{\emptyset} + 1 = \emptyset$.

But this is equivalent to $\dfrac{1}{\emptyset} + \dfrac{1}{1} = \dfrac{1}{\emptyset^{-1}}$, which, in turn, is equivalent to $\dfrac{1}{\emptyset^1} + \dfrac{1}{\emptyset^0} = \dfrac{1}{\emptyset^{-1}}$, and the statement is proved for $m=1$.

(b') Assume true for $m=k$. That is, $\dfrac{1}{\emptyset^k} + \dfrac{1}{\emptyset^{k-1}} = \dfrac{1}{\emptyset^{k-2}}$.

(c') Prove true for $m = k+1$.

Since $\dfrac{1}{\emptyset^k} + \dfrac{1}{\emptyset^{k-1}} = \dfrac{1}{\emptyset^{k-2}}$, then

$$\frac{1}{\emptyset}\left(\frac{1}{\emptyset^k} + \frac{1}{\emptyset^{k-1}}\right) = \left(\frac{1}{\emptyset^{k-2}}\right)\frac{1}{\emptyset}, \text{ or}$$

$$\frac{1}{\emptyset^{k+1}} + \frac{1}{\emptyset^k} = \frac{1}{\emptyset^{k-1}}.$$

Therefore true for $m = k + 1$, and by the principle of mathematical induction, the statement holds for all positive integral values of m. Since $m = -n$, we have $\dfrac{1}{\emptyset^{-n}} + \dfrac{1}{\emptyset^{-n-1}} = \dfrac{1}{\emptyset^{-n-2}}$ or $\emptyset^n + \emptyset^{n+1} = \emptyset^{n+2}$, and the theorem is proved for all integral values of n.

The Relationship between Cubit Measure and Phi

Royal Cubit $\approx .5236 \approx (\emptyset^2 + \emptyset + 1)(.1)$.

Little cubit $\approx .4236 \approx (\emptyset^2 + \emptyset)(.1)$.

In each case the unit is taken to be .1 meter.

Proof of the Construction of the Golden Cut of a Line Segment

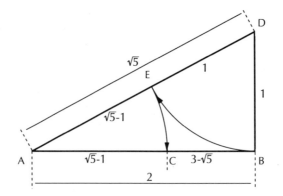

If we let AB = 2 and construct the perpendicular at B such that BD = 1, by the Pythagorean Theorem , AD = $\sqrt{5}$. Point E divides $\overline{AD}$ into segments AE and ED having lengths $\sqrt{5}$ - 1 and 1, respectively. Then point C divides $\overline{AB}$ into segments AC and CB having lengths $\sqrt{5}$ - 1 and 2 - ($\sqrt{5}$ - 1) = 3 - $\sqrt{5}$, respectively.

Comparing AB to AC we have

$$\frac{2}{\sqrt{5}-1} \cdot \frac{\sqrt{5}+1}{\sqrt{5}+1} = \frac{2+2\sqrt{5}}{5-1} = \frac{2+2\sqrt{5}}{4} = \frac{1+\sqrt{5}}{2}$$
$$= \emptyset.$$

Also, $\dfrac{AC}{CB} = \dfrac{\sqrt{5}-1}{3-\sqrt{5}} = \dfrac{-1+\sqrt{5}}{3-\sqrt{5}} \cdot \dfrac{3+\sqrt{5}}{3+\sqrt{5}}$

$$= \frac{-3+2\sqrt{5}+5}{9-5} = \frac{2+2\sqrt{5}}{4} = \frac{1+\sqrt{5}}{2} = \emptyset.$$

Therefore, C is the Golden Cut of $\overline{AB}$.

Lengths of the Dynamic Rectangles and Their Reciprocals*

Name	Length	Width of Reciprocal
$\sqrt{2}$ Rectangle	1.4142	.7071
$\sqrt{3}$ Rectangle	1.7321	.5773
$\sqrt{4}$ Rectangle	2.0000	.5000
$\sqrt{5}$ Rectangle	2.2361	.4472
Golden Rectangle	1.6180	.6180
$\sqrt{\emptyset}$ Rectangle	1.2720	.7862
$\emptyset$+1 Rectangle	2.6180	.3820

*rounded to the nearest ten thousandth
In each of the rectangles the width is taken to be 1.

Proof of the Construction of the Dynamic Rectangles

Given square ABCD with sides of length 1, prove the resulting rectangles are Dynamic or Root Rectangles.

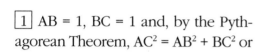

1 AB = 1, BC = 1 and, by the Pythagorean Theorem, $AC^2 = AB^2 + BC^2$ or

$AC^2 = 1^2 + 1^2 = 1 + 1 = 2$, or $AC = \sqrt{2}$

E and C are equidistant from A.

Therefore, $AE = \sqrt{2}$, and AEFD is a $\sqrt{2}$ Rectangle.

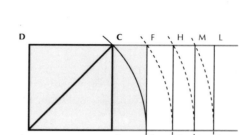

2 $AE = \sqrt{2}$, EF = 1 and $AF^2 = AE^2 + EF^2$.

$AF^2 = \sqrt{2}^2 + 1^2 = 2 + 1 = 3$, so $AF = \sqrt{3}$.

F and G are equidistant from A.

Therefore, $AG = \sqrt{3}$ and AGHD is a $\sqrt{3}$ Rectangle.

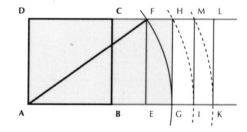

3 $AG = \sqrt{3}$, GH = 1 and $AH^2 = AG^2 + GH^2$.

$AH^2 = \sqrt{3}^2 + 1^2 = 3 + 1 = 4$, so $AH = \sqrt{4}$.

H and I are equidistant from A.

Therefore, $AI = \sqrt{4}$ and AIMD is a $\sqrt{4}$ Rectangle.

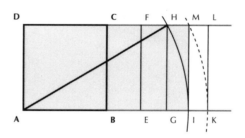

4 $AI = \sqrt{4}$, IM = 1 and $AM^2 = AI^2 + IM^2$.

$AM^2 = \sqrt{4}^2 + 1^2 = 4 + 1 = 5$, so $AM = \sqrt{5}$.

M and K are equidistant from A.

Therefore, $AK = \sqrt{5}$, and AKLD is a $\sqrt{5}$ Rectangle.

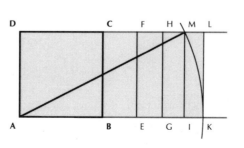

Proofs of Some Properties of the Fibonacci Numbers

$\boxed{1}$ The sum of the first n terms of the Fibonacci sequence is 1 less than the second term beyond F_n, or $F_1 + F_2 + F_3 + ... + F_n = F_{n+2} - 1$.

Proof: Since the Fibonacci sequence is a summation sequence, $F_n + F_{n+1} = F_{n+2}$ and, therefore, $F_n = F_{n+2} - F_{n+1}$.

$$F_1 = F_3 - F_2 = F_3 - 1$$
$$F_2 = F_4 - F_3$$
$$F_3 = F_5 - F_4$$
$$\vdots$$
$$F_{n-1} = F_{n+1} - F_n$$
$$F_n = F_{n+2} - F_{n+1}$$

By adding both sides of the equations we have:

$$F_1 + F_2 + F_3 + ... + F_{n-1} + F_n = F_{n+2} - F_{n+1} + F_{n+1} - ... -F_4 + F_4 - F_3 + F_3 - 1.$$
$$\text{Or } F_1 + F_2 + F_3 + ... + F_n = F_{n+2} - 1.$$

$\boxed{2}$ The sum of the even numbered terms of the Fibonacci sequence through F_{2n} is one less than F_{2n+1}, or $F_2 + F_4 + F_6 + ... + F_{2n} = F_{2n+1} - 1$.

Proof: $F_n + F_{n+1} = F_{n+2}$ or $F_{n+1} = F_{n+2} - F_n$.

$$F_2 = F_3 - F_1 = F_3 - 1$$
$$F_4 = F_5 - F_3$$
$$F_6 = F_7 - F_5$$
$$\vdots$$
$$F_{2n-2} = F_{2n-1} - F_{2n-3}$$
$$F_{2n} = F_{2n+1} - F_{2n-1}$$

Adding we get:

$$F_2 + F_4 + F_6 + ... + F_{2n-2} + F_{2n} = F_{2n+1} - F_{2n-1} + F_{2n-1} - ... - F_5 + F_5 - F_3 + F_3 - 1.$$
$$\text{Or } F_2 + F_4 + F_6 + ... + F_{2n} = F_{2n+1} - 1.$$

$\boxed{3}$ The sum of the odd numbered terms of the Fibonacci sequence through F_{2n-1} is the next term, F_{2n}, or $F_1 + F_3 + F_5 + ... F_{2n-1} = F_{2n}$.

Proof:

$$F_1 = 1$$
$$F_3 = F_4 - F_2 = F_4 - 1$$
$$F_5 = F_6 - F_4$$
$$\vdots$$
$$F_{2n-3} = F_{2n-2} - F_{2n-4}$$
$$F_{2n-1} = F_{2n} - F_{2n-2}$$

Adding we get:

$$F_1 + F_3 + F_5 = \ldots + F_{2n-3} + F_{2n-1} = F_{2n} - F_{2n-2} + F_{2n-2} - \ldots - F_4 + F_4 - 1 + 1.$$
Or $F_1 + F_3 + F_5 + \ldots + F_{2n-1} = F_{2n}$.

$\boxed{4}$ The square of a Fibonacci number and the product of the Fibonacci numbers that precede and follow it in the sequence differ by one,
   or $(F_n)^2 - (F_{n-1})(F_{n+1}) = \pm 1$.

Proof: We can adjust the subscripts without altering the sense of the statement to read $(F_{n+1})^2 = (F_n)(F_{n+2}) + (-1)^n$ where the exponent n indicates whether the squared term is one more than the product or one less than the product, since $(-1)^n = 1$ if n is even and $(-1)^n = -1$ if n is odd.

This property of the Fibonacci sequence can be proved by mathematical induction.

(a) Prove the statement true for n = 1.
$$(F_{1+1})^2 = (F_2)^2 = 1^2 = 1.$$
$$(F_1)(F_{1+2}) + (-1)^1 = (F_1)(F_3) - 1 = (1)(2) - 1 = 2 - 1 = 1.$$
Therefore, $(F_{1+1})^2 = (F_1)(F_{1+2}) + (-1)^1$, and the statement is true for n = 1.

(b) Assume the statement true for n = k.
   Then $(F_{k+1})^2 = (F_k)(F_{k+2}) + (-1)^k$.

(c) Prove true for n = k+1.
By adding $(F_{k+1})(F_{k+2})$ to both sides of the equation in (b) we get
$$(F_{k+1})^2 + (F_{k+1})(F_{k+2}) = (F_k)(F_{k+2}) + (F_{k+1})(F_{k+2}) + (-1)^k.$$
 Or, by factoring,
$$F_{k+1}(F_{k+1} + F_{k+2}) = F_{k+2}(F_k + F_{k+1}) + (-1)^k.$$
But $F_{k+1} + F_{k+2} = F_{k+3}$ and $F_k + F_{k+1} = F_{k+2}$, so we can substitute to get
$$(F_{k+1})(F_{k+3}) = (F_{k+2})^2 + (-1)^k \quad \text{or} \quad (F_{k+2})^2 = (F_{k+1})(F_{k+3}) + (-1)^{k+1}.$$

Therefore, the statement is true for n = k+1 and, by the principle of mathematical induction, is then true for all integral values of n.

C Geometric Formulas

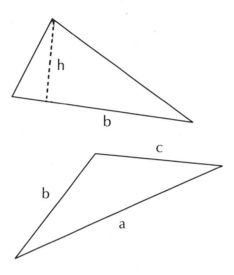

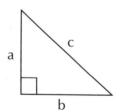

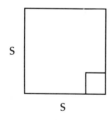

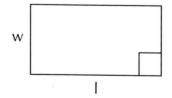

Triangle

Area: $A = \frac{1}{2}bh$ where b is the measure of any side and h is the measure of the altitude to that side.

$A = \sqrt{s(s-a)(s-b)(s-c)}$ where a, b and c are the lengths of the sides and $s = \frac{1}{2}(a+b+c)$.

Right Triangle

Pythagorean Theorem: $a^2 + b^2 = c^2$ where a and b are the measures of the legs and c is the measure of the hypotenuse.

Square

Area: $A = s^2$
Perimeter: $P = 4s$ where s is the measure of a side.

Rectangle

Area: $A = lw$
Perimeter: $P = 2l + 2w$ where l is the measure of the length and w is the measure of the width.

260

Parallelogram

Area: A = bh where *b* is the measure of a side and *h* is the measure of an altitude to that side.

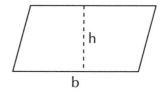

Rhombus

Area: $A = \frac{1}{2}ab$ where *a* and *b* are the lengths of the diagonals.

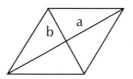

Trapezoid

Area: $A = \frac{1}{2}h(a + b)$ where *a* and *b* are the measures of the parallel sides and *h* is the distance between those sides.

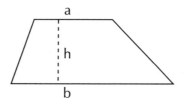

Circle

Area: $A = \pi r^2$

Circumference: $C = \pi d = 2\pi r$ where *r* and *d* are the radius and diameter, respectively, and π may be approximated by 3.14 or 22/7.

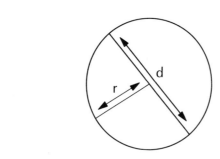

Regular Polygon

Area: $A = \frac{1}{2}asn$

Perimeter: P = sn where *a* is the measure of the apothem, *s* is the measure of a side, and *n* is the number of sides.

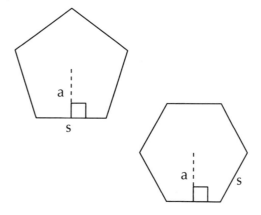

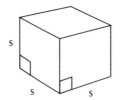

Cube

Volume: $V = s^3$

Surface Area: $S = 6s^2$ where s is the measure of a side.

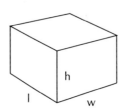

Rectangular Solid

Volume: $V = lwh$

Surface Area: $S = 2lw + 2wh + 2lh$ where l is the length, w is the width, and h is the height.

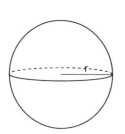

Sphere

Volume: $V = \frac{4}{3}\pi r^3$

Surface Area: $S = 4\pi r^2$ where r is the radius and π may be approximated by 3.14 or 22/7.

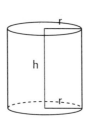

Right Circular Cylinder

Volume: $V = \pi r^2 h$

Lateral Surface Area: $L = 2\pi rh$ (does not include the area of the circular ends).

Total Surface Area: $S = 2\pi rh + 2\pi r^2$ where r is the radius, h is the height, and π may be approximated as above.

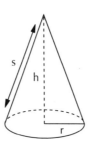

Right Circular Cone

Volume: $V = \frac{1}{3}\pi r^2 h$

Lateral Surface Area: $L = \pi rs$ (does not include the area of the circular base).

Total Surface Area: $S = \pi r^2 + \pi rs$

Slant Height: $s = \sqrt{r^2 + h^2}$

In each case, r is the radius, h is the height, s is the slant height, and π is as above.

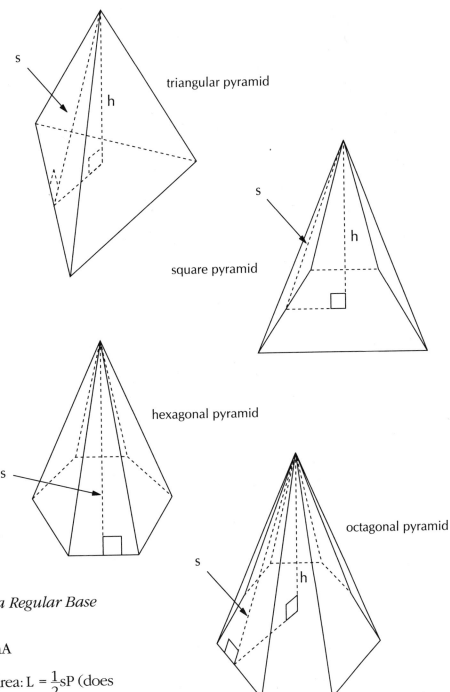

triangular pyramid

square pyramid

hexagonal pyramid

octagonal pyramid

Right Pyramid on a Regular Base

Volume: $V = \frac{1}{3}hA$

Lateral Surface Area: $L = \frac{1}{2}sP$ (does not include the area of the base).

Total Surface Area: $S = \frac{1}{2}sP + A$

A is the area of the base (choose the appropriate formula for the shape of the base), *P* is the perimeter of the base (found by multiplying the length of a side by the number of sides), *h* is the height, and *s* is the slant height (the altitude from the base of one of the triangular faces).

D Art Materials &Techniques

Contents

Approaching an Art Project

If doing artwork is new to you, the following steps may help you on your journey.

1 Decide upon the subject matter of the work, which might be still life, portraiture, landscape or pure design. What elements do you find most interesting about this particular subject matter? What is most important to you about this subject? What qualities do you want to emphasize?

2 Choose the size, shape and proportions of your two-dimensional support, which might be canvas, paper, or board. Do this carefully. The outer proportions will determine how successfully you can subdivide the interior surface-space. Certain proportions create dynamic and harmonic divisions, while others do not subdivide well.

Be concerned with size. Some ideas are more intimate and require smaller dimensions. Some ideas are large enough for murals. In some ways, it is analogous to the differences between a poem, a short story or a novel.

3 Work out the arrangement and placement of the large shapes on an armature. Do not consider the specific details at this time. They will come later. Do a group of quick sketches for potential arrangements of forms. Choose at least three ideas from the sketches. Refine them in a slightly larger size. This might be the place to experiment with different materials. Keep in mind the proportions of your original format and consider the following:

a. Will you be working with a color structure or only with black and white?

b. Are you working only with a strong dark and light contrast?

c. Are you going to have a dark, a light and two or three other value changes?

d. Are you going to work with only 3 - 5 light values?

e. Are you going to work with only 3 - 5 dark values?

In your studies, either find your values by your choice of papers, or create them by covering sheets of paper with paint, colored pencils, markers or inks in the values of your choice. Then cut them into the desired shapes and glue onto the surface. Consider the other options of texture and pattern.

4 Once the values and placement of key shapes are established, block them onto your final composition.

5 On top of this understructure, you will be working in your basic colors and adding your details. Decide where your center of interest will be and work to bring all the color, movement and rhythms to that place. Pay attention to detail and craftsmanship.

6 Bon voyage!

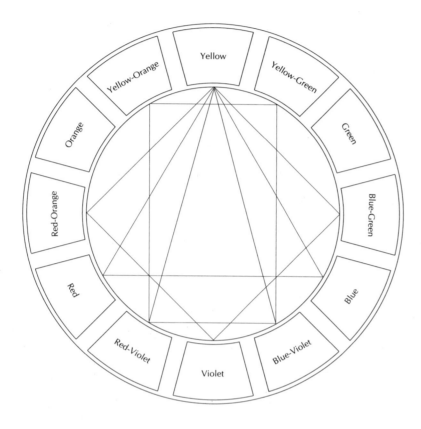

Color Wheel

The color wheel is based on a twelve tone structure. Yellow, blue and red are the three *primary* colors. Theoretically, all the other colors can be mixed from only these three. It is best, however, when purchasing pigments to include white, black, and violets to gain the maximum amount of flexibility in mixing color.

A *secondary* color is obtained by mixing a pair of primary colors. The secondary colors are orange, green and violet.

A mixture of a primary and a secondary color yields one of the *tertiary* colors, which are the remaining ones on the wheel.

Tints are obtained by mixing the color with white, while *tones* are gotten by mixing gray into a color, and *shades* are obtained by mixing in black.

In this particular color wheel, the geometric figures in the center circle provide different color structures. They can be placed anywhere within the circle and the vertices will indicate which colors will work together. The two triangles give a triadic color structure (one made up of three colors) while the rectangles give tetradic structures (made up of four colors).

Do not always rely on given systems, however. Be experimental in the mixing and placing of colors.

Color Harmonies

In order for a color scheme to be harmonious, a few basic guidelines could be followed in limiting the number of colors used. This can be done by using one of the following schemes:

Monochromatic

Uses one color and the light and dark variations thereof.

Complementary

Uses a pair of colors directly opposite one another with light and dark variations thereof.

Analogous

Uses three adjacent colors with light and dark variations thereof.

Triad

Uses three colors that form the vertices of either an isosceles or equilateral triangle with the light and dark variations thereof.

Tetrad

Uses four colors that form the vertices of a rectangle (including the square) with the light and dark variations thereof.

Basic Materials

The following materials are suggested if you are just beginning your investigation of art. This list is by no means complete as the modern market abounds in art supplies of every description. It is, rather, a list of basic supplies, relatively easy to use, offered with the hope that your first attempts at artistic endeavors will be successful and will encourage you to continue.

Adhesives

See Adhesive Chart on page 268.

Brushes

Brushes are constructed for use with different media. There is a synthetic type designed for acrylic polymer paints. Which type will depend upon the consistency of the paint used and the effect desired. Have at least three different sizes, ranging from fine for detail work to 1/2" for fill-in work.

Construction Tools

Metal straightedge

Compass with opening wide enough for constructing good sized circles

Scissors

Utility knife with extra blades

T-square and triangles

Fabric Crayons

These can be used to draw original designs on paper which can then be heat transferred by iron to fabric. The design is reversed when transferred.

Inks

These come in transparent, opaque and metallic permanent types. They can be used in combination with other materials.

Markers

These come in a great many varieties from waterbase to permanent (this only refers to the markers being waterproof, not lightfast); from transparent to opaque; from fluorescent to metallic. Test before buying in quantity.

Paints

Gouache—waterbase opaque paint with an essentially matte finish, not necessarily permanent.

Transparent watercolor—waterbase paint which is not permanent. Allows the white of the paper to show through as a visual element.

Acrylic—waterbase but permanent when dry.

Include: (2 oz. tubes or jars)
white   yellow   black   blue
red   violet or purple
and medium (matte or gloss type). This paint can be used either transparently or opaquely, depending on the desired effect. Opacity may require between three and five coats. Brushes should be kept in water to prevent the paint from hardening and ruining them. This can happen very quickly since the paint dries fast.

Paper

See Paper Chart on page 269.

Pencils

Graphite pencils
H = hard   B = soft The number and the letter on the side of the pencil indicate degree of hardness or softness. HB is a middle range.

Colored pencils
These come in a variety of types. Some can be erased, some are water soluble and can be used like watercolor, some are oil base. Buy a good quality. They can be purchased individually or in sets, in hard or soft types, and in stick form. They are blended by drawing on top of one another. To obtain clean, precise lines and to reduce smudging, it would be best to avoid crayons, oil pastels, pastels, and charcoal.

Adhesives and Their Uses

A	Contact Cement	F	Resin Glue
B	Epoxy Cement	G	Rubber Cement Glue
C	Epoxy Metal Cement	H	Waterproof Glue
D	Plastic Resin Urea Base	I	White Glue
E	Polymer Medium (Acrylic Emulsion)	N	Nonporous Material
		P	Porous Material

MATERIALS TO BE GLUED	A	B	C	D	E	F	G	H	I	N/P
China and glass, ceramic tiles	x									x
Collagraph items for printmaking			x				x			x
Cloth with wood					x	x		x		x
Leather and wood		x			x	x		x		x
Metal to metal	x	x	x						x	
Metal to wood	x	x	x					x	x	
Paper and cloth					x	x		x		x
Paper to paper and cardboard					x	x	x	x		x
Plastic foams (Styrofoam)		x		x	x	x		x	x	x
Plastics and Vinyls				x	x	x			x	
Plastic to wood			x		x	x			x	
Rubber							x	x	x	
Rubber to wood or metal	x	x	x						x	
Stone & concrete to other items	x								x	
Sand, stones, beads				x	x	x		x	x	
Wood outdoors				x			x			x
Wood to wood				x		x		x	x	x

General Comments:

Labels should be checked for proper use of the product.

When gluing dissimilar materials, their porosity should be determined. When combining two types of materials, the type recommended for the non-porous material should be used.

Duco Cement or LePage's Household Cement are good all-purpose adhesives. LePage's Glue is good for general paper work.

Shoe Patch, purchased at a sporting goods store, is a very strong adhesive that might be tried when others fail.

Polymer medium is useful for preserving, glazing and sealing. It comes in both matte and gloss finish.

With the exception of polymer medium, available at art supply stores, all other adhesives can be found in hardware stores.

Adhesives should be tested on a small sample first.

Paper Adhesives	Characteristics	Uses
Glue stick	Very tacky, very strong	Adhering tabs and small areas
Metylan cellulose paste	Inexpensive, dissolves in water	Paper maché, paper, cardboard
Mucillage	Amber colored, tacky, syrupy	Paper
Polyvinyl acetate (Elmer's, Sobo, Duratite, etc.)	White, dries clear, strong	Decoupage, paper maché
Plastic glaze (Modge Podge, Art Podge)	White, dries clear, strong	Protecting paper surfaces
Rubber Cement	Dries quickly, excess easily rubbed away	One coat temporary, two coats permanent

Papers and Their Uses

TYPE	WEIGHT	CHARACTERISTICS	TEXTURE & COLOR	USES
Albanene	Light	100% rag, transparent easily erasable, will not yellow or crack	White, smooth	Tracing
Bristol board	Light to heavy	Tough, many thicknesses	Very smooth, white	Acrylic studies, sculpture
Bond	Light	Somewhat fragile	White, smooth	Pencil studies
Carbon transfer	Very light	Carbon on one side	Several colors	Transfer
Construction	Medium	Inexpensive, fades easily	Full range of color	All-purpose, temporary
Graphite	Light	One side coated with graphite	Gray	Transfers to board & paper
Illustration board	Light to heavy	Lies flat, consistent working surface	White, grays, blacks smooth and rough surfaces	Polyhedra models, acrylic studies
Craft	Medium	Strong	Brown	All purpose
Metal foil	Light	Paper backed with foil	Varied color, shiny and matte	Sculpture
Oak tag	Heavy	Strong, folds without cracking	Full range of colors, smooth	Polyhedra models, paper sculpture
Press-apply	Light	Adhesive backed	Smooth, intense rainbow colors	Surface decoration
Scorasculpture (Dennison Co.)	Medium	Easily scored, very workable	Smooth, white	Polyhedra models, paper sculpture
Tissue	Very light	Translucent	Full range of color, plain or variegated	Surface decoration
Velour	Medium	Fragile surface	Full range of color Flocked surface	Surface decoration

Materials & Techniques for Polyhedra Construction

Tools

Compass

Metal Straightedge

Utility knife or X-acto Knife with #11 blade

Hole punch (optional)

Bobby pins or various sized paper clips, binder clamps or paper clamps for temporarily holding joints together.

Cutting board of either cardboard, chipboard, masonite or the more expensive self-healing cutting surface bought at art or fabric stores or architectural supply firms.

Tweezers

Materials

Pencils

Surface material—some suggestions follow:

a. Tag board, card stock or colored poster board (sometimes called railroad board),

b. Two-ply bristol board or two-ply strathmore paper (both take color well),

c. Clear or colored acetate or rigid vinyl sheet plastic— transparent or opaque,

d. Balsa wood in sheet or stick form.

Adhesive—see chart on page 268.

Tapes

a. Scotch brand Magic Tape #810, half inch for paper models,

b. Scotch Brand Polyester Tape #853, half inch for plastic models,

c. Architect's tape—comes in a variety of widths and colors—decorative for joining edges.

Color for Surfaces

a. Permanent ink markers,

b. Opaque ink markers,

c. Acrylic or alkyd paints,

d. Polymer matte or gloss medium.

Procedure for Construction

1. Obtain the appropriate materials and supplies. Allow plenty of time and space to work slowly, neatly, and accurately.

2. Choose a specific polyhedron and construct the template or plane "net" for the polyhedral form. Take care with this construction or the inaccuracies will be compounded as you construct additional units.

3. A convenient working size is a 2" or 3" edge length. Make sure the edges of all the faces are of equal length. The larger the polyhedron, the sturdier the material must be.

4. Cut the units or net on a cutting board.

5. Some methods of construction require tabs on the individual units or on the nets; some do not. When working with rigid plastic, the units are taped together and do not require tabs.

6. Bond edges or tab faces carefully, making sure that the joints are accurate and tight.

7. On lightweight paper, use a ballpoint pen to score all the necessary fold lines; otherwise, use the x-acto knife without cutting through.

8. Paint may be applied before scoring or bending into three dimensions. Or the form may be constructed before the paint is applied. Use multiple coats of paint for a good surface.

9. Use a matte or gloss varnish over the paint for a final protective coat.

E Answers to Selected Problems

It is highly recommended that you do each of the constructions in a given chapter before attempting the problems, some of which require knowledge of the constructions. For quite a few of the problems, there are many "right" answers. Therefore, much of this section consists of hints or directions on how to start.

Chapter 1

1. Draw a line segment and mark off a segment congruent to *a*. At one endpoint, swing an arc with radius equal to *b*, and at the other swing an arc with radius equal to *c*. Draw the segments from the endpoints of the segment to the point of intersection of the two arcs.

2. This problem is done the same way as Problem 1. The only difference is that the given segments already form a triangle. The compass is again used to measure the lengths of the sides.

3. a. Since 8 is a power of 2, this part can be done by repeated bisection. See Fig. 1.1, and carry the process one step further to bisect each of the four congruent segments.

   b. Since 7 is not a power of 2, this must be done by the method of Construction 9.

4. Since an altitude is a segment from a vertex of a triangle perpendicular to the opposite side (see Glossary for diagram), this problem is asking you to construct perpendiculars from A to side BC, B to side AC, and C to side AB. Use the method of Construction 5 to construct a perpendicular to a line through a point not on the line. Note: you may extend the sides of the triangle for ease of construction. Remember that you can rotate your paper if you prefer to work with the figure oriented the same way as in Construction 6 (the point "above" the line). All three altitudes of a triangle intersect in a single point. This fact allows you to check the accuracy of your construction.

5. A rectangle has four sides in which opposite pairs are parallel and congruent. Each angle is a right angle. Since perpendicular lines form right angles, begin by drawing a line and choosing an arbitrary point through which to construct a perpendicular (as in Construction 4). The rest of the rectangle can be constructed simply by measuring with the compass to get two sets of congruent segments.

6. Since all three angle bisectors meet in a single point, point P will be the same regardless of which two angles are bisected. Point P determines the center of the *inscribed circle* (see Glossary for diagram), which is tangent to each side of the triangle.

7. The point of intersection of the perpendicular bisectors of the sides of a triangle (again, all three will meet in

a single point) is the center of the circle circumscribed about the triangle. The triangle will be *inscribed* in the circle (see Glossary), and all three vertices of the triangle will lie on the circle.

8. Begin with a line segment of arbitrary length and at one endpoint construct an angle congruent to angle A. At the other endpoint construct an angle congruent to angle B. Extend the sides of the angles until they intersect. A different sized triangle can be gotten by altering the length of the initial segment.

9. Extend line AB to make the construction easier. Again, remember that you can rotate your paper to get the same orientation as in Construction 7.

Chapter 2

1. $\frac{B}{A}$ means divide the answer to part B by the answer to part A. In this case, the number obtained in part A becomes the unit. The closer the answer is to 1.6, the closer you are to Greek perfection! Using metric measure makes this problem easier.

4. Use the technique of Construction 10.

5. Use the technique of Construction 11.

6. a. Use Construction 13 to obtain a regular pentagon, and then construct its pentagram.

b. Begin with Construction 14.

7. The rectangles are close to Golden Rectangles if the result of dividing length by width is close to 1.6. Use metric units of measure to make this problem easier.

8. Use the technique of Construction 15 to obtain a Golden Rectangle. The process may be repeated beginning with different sized squares, or similar rectangles may be constructed using the technique of Construction 17.

9. Begin with a regular pentagon and find Golden Cuts of the sides.

Chapter 3

1. "Line thickness" should not be taken into consideration as a part of these constructions. In each case, begin by finding the midpoints of the sides of an equilateral triangle.

5. Check Appendix C for the formula for for the area of a triangle. You will have to devise your own unit of measure for this problem.

6. Remember that the angles of a triangle always add up to 180°. Use the chart on page 93 to see, in each case, which special triangle is closest to your result. Each triangle should have at least two congruent sides, and therefore, will either be isosceles or equilateral. If isosceles, which special type is it?

Chapter 4

3. Assume the sides are divided at Golden Cuts.

4. Begin by isolating a rectangle and constructing its reciprocal.

5. After dividing the length by the width, use the information on page 256 to determine whether the rectangle is close to one described in this chapter.

Chapter 5

2. "Left over" portions in each case may be disregarded.

4. Use Construction 40 to subdivide the sides of the polygons. See Fig. 5.6 for inspiration.

Chapter 6

2. See the chart on Angular Spirals, page 209.

3. Use the technique of Construction 42.

4. Use Construction 43.

5. Be sure to extend the sides of the

hexagon in one direction only. If the sides are extended to the left (looking at the sides from the interior of the figure) as in Construction 44, the spiral will curve in a clockwise direction. If extended to the right, the spiral will turn in a counterclockwise direction.

7. Use the technique of Construction 45.

8. See the description of and technique for obtaining the Dynamic Rectangles Spiral on page 233.

Glossary

A

acute angle An angle that measures between 0° and 90°.

acute angles

acute triangle A triangle having three acute angles.

acute triangles

adjacent Next to.

altitude *In a triangle:* the line segment from a vertex perpendicular to the opposite side.

Every triangle has three altitudes which might not all be in the interior of the triangle.

In an acute triangle all three altitudes lie in the interior of the triangle.

altitudes: $\overline{AF}, \overline{BD}, \overline{CE}$

In a right triangle the legs are also altitudes.

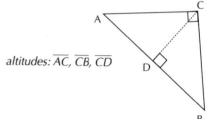

altitudes: $\overline{AC}, \overline{CB}, \overline{CD}$

In an obtuse triangle two of the altitudes lie in the exterior of the triangle.

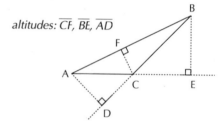

altitudes: $\overline{CF}, \overline{BE}, \overline{AD}$

In a parallelogram: the altitude is the distance between two parallel sides. Any line segment joining two sides and perpendicular to both is *an* altitude.

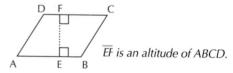

$\overline{EF}$ *is an altitude of ABCD.*

analogous color scheme A color scheme that uses hues that are next to each other on the color wheel. Light and dark values of the hues may be included.

angle The figure formed by two rays having a common endpoint. The common endpoint is called the vertex of the angle, and the rays are called the sides of the angle.

∠ABC names the angle whose vertex is B such that A is a point on one side and C is a point on the other. This angle could also

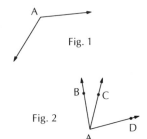

Fig. 1

Fig. 2

be named ∠CBA. The vertex B must be between the other two points named.

If there is no ambiguity, an angle may be named only by its vertex. ∠A makes sense in Fig. 1 but not in Fig. 2 since there are three angles at A, namely ∠BAC, ∠CAD, and ∠BAD.

apothem   A line segment from the center of a circle perpendicular to a side of an inscribed regular polygon.

$\overline{OF}$ is an apothem of polygon ABCDEF.

arc   The figure formed by two points and the portion of a circle between those two points.

The two points are called the endpoints of the arc, and an arc is named by its endpoints.

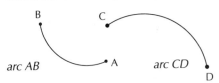

arc AB   arc CD

If two points lie on a circle, they determine two different arcs. To avoid confusion, the arcs may be named by the endpoints and any other point on the arc. The third point is listed between the endpoints. In the first circle, A and B are endpoints of a diameter. Therefore, arc ACB and arc ADB are both semicircles. In the second, A and B are

not endpoints of a diameter. Arc ACB is called the minor arc (the smaller one) and arc ADB is called the major arc (the larger one).

Archimedean spiral   See spiral.

area   The measure in square units of any bounded portion of a plane.

arithmetic sequence   A sequence in which each term is gotten by adding the same number to the preceding term.

eg. 1, 5, 9, 13, 17, ...is an arithmetic sequence where 4 is added to each term to produce the following term.

axis   A line used as a reference.

In the Cartesian coordinate plane all points are located with reference to two perpendicular axes: the x axis and the y axis.

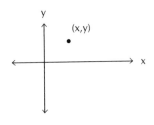

An axis of symmetry is a line which divides the plane into two halves which are mirror images of each other (also called an axis of reflection or a mirror).

B

Baravelle Spiral   The straightline design based on repeatedly connecting consecutive midpoints of sides of regular polygons and shading to enhance the structure.

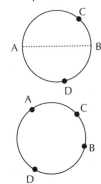

base angles *In an isosceles triangle:* the two congruent angles.

∠A and ∠B are the base angles of △ABC, and ∠R and ∠S are the base angles of △RST.

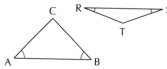

In a trapezoid: a pair of angles with a base (one of the parallel sides) as a common side. Each trapezoid has two pairs of base angles.

In ABCD, ∠A and ∠B are one pair of base angles, and ∠C and ∠D are the other pair.

bisect To divide into two congruent figures.

Line l bisects $\overline{PQ}$.

$\overrightarrow{BD}$ *bisects ∠ABC.*

bisector The line or ray that divides a segment or angle into congruent figures.

bisector, perpendicular A line that passes through the midpoint of a segment and is at right angles to the segment.

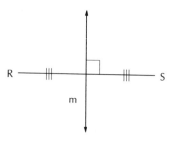

Line m is the perpendicular bisector of $\overline{RS}$.

C

Cartesian coordinate plane A rectangular system for locating points in a plane by using a pair of perpendicular axes: the horizontal line is the x axis and the vertical line is the y axis.

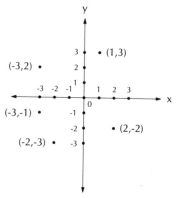

A point is located by an ordered pair of numbers (x,y) which places it with respect to both axes.

cella In Greek temples, the rectangular interior space which housed the cult statue.

center An point from which all points on a circle are equidistant.

P is the center of the circle.

central angle The angle formed by two radii of a circle.

In circle P, ∠APB is a central angle.

chord A line segment whose endpoints lie on a circle.

In circle P, $\overline{RS}$ is a chord.

circle The set of all points in a plane equidistant from a given point.

Although it is not part of the circle, a circle is named by its center. Circle P is a circle whose center is P (see center).

circumference The distance around a circle.

The circumference of a circle is given by the formula C = πd, where d is the diameter of the circle.

coefficient A factor in a product generally taken to mean a number multiplied by a variable quantity.

In the expressions 2x, -3a²b, -pq: 2 is the coefficient of x, -3 is the coefficient of a²b, and -1 is the coefficient of pq.

collinear Lying on the same line.

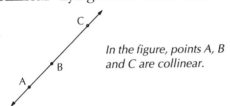

In the figure, points A, B and C are collinear.

commensurable Able to be measured by a common unit.

compass An instrument used for drawing circles and arcs.

complementary angles Two angles the sum of whose measures is 90°.

The angles are said to be complements of each other. ∠ ABC is complementary to ∠ CBD. Angles P and Q are complementary angles.

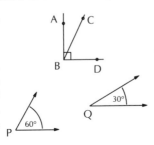

complementary color scheme A color scheme that uses two hues that are directly opposite each other on the color wheel. It can include light and dark values of the hues.

composite number A natural number, different from one, that is not a prime.

The number one is neither prime nor composite.

concave polygon A polygon that has at least one interior angle that measures more than 180°.

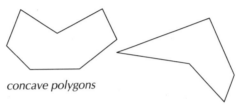

concave polygons

concentric Having the same center.

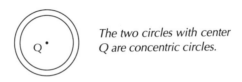

The two circles with center Q are concentric circles.

concho spiral The three-dimensional equivalent of the logarithmic spiral.

This is the spiral found on the surfaces of sea shells and pine cones.

concurrent Occurring at the same time.

cone A three-dimensional figure with a circular base and a surface generated by the lines passing through a point outside the plane of the circle and the points on the circle.

See Appendix C for formulas related to the cone.

Circle Q defines the base of the cone. P is called the vertex of the cone. If PQ is perpendicular to the plane of the circle, the cone is a right circular cone.

congruent Having the same size and shape.

conical helix The spiral-like curve formed by the uniform wrapping of a line around a cone from its vertex to its base.

consecutive In order or in sequence.

eg. 2, 3, 4 are consecutive integers. 5 and 7 are consecutive odd integers.

In a polygon, consecutive sides or consecutive vertices are those that appear in order as you travel around the polygon in either direction, starting at any vertex.

convex polygon A polygon whose interior angles each measure less than 180°.

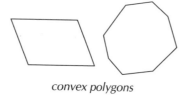

convex polygons

cube *As a noun:* A polyhedron having six square faces.

As a verb: To raise to the third power.
eg. $4^3 = 4$ cubed $= 4 \times 4 \times 4 = 64$.

cubit An ancient Egyptian unit of linear measure.

The little cubit measured .4236 m.
The Royal Cubit measured .5236 m.

curve A continuous set of points having only one dimension, namely length.
Note: by this definition, a straight line is a curve.

cylinder A three-dimensional figure generated by a circle whose center moves along a line segment.

See Appendix C for formulas related to the cylinder.

Circles P and Q define the bases of the cylinder. If $\overline{PQ}$ is perpendicular to the planes of circles P and Q, then the cylinder is a right circular cylinder.

D

decagon A polygon having ten sides.

decagon

deduction See deductive reasoning.

deductive reasoning A method of reasoning that allows specific conclusions to be drawn from general truths. This is the form of reasoning used in mathematical proof.

degree (°) The unit of measure for angles.
There are 360° in a circle.

denominator In a fraction, the number that is written below the line.

diagonal A line segment that joins the nonconsecutive vertices of a polygon.

The number of diagonals a polygon has is dependent on the number of sides.

Each of the line segments in the interiors of the figures is a diagonal.

diameter A chord that passes through the center of a circle.

A circle has infinitely many diameters — all the

same length. The measure of any one of them is called the diameter of the circle. $\overline{QR}$ is a diameter of circle P. QR is the diameter of circle P. A diameter cuts the circle into two semicircles.

$$\frac{AB}{AC} = \frac{AC}{CB}$$

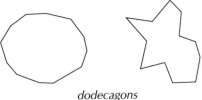

$$\frac{a+b}{a} = \frac{a}{b}$$

distance *Between points:* the length of the line segment joining the points.

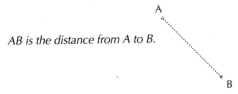

AB is the distance from A to B.

dodecagon A polygon having twelve sides.

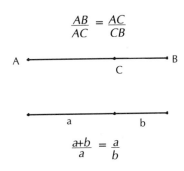

dodecagons

From a point to a line: the length of the line segment from the point to the line and perpendicular to the line.

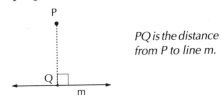

PQ is the distance from P to line m.

dodecahedron A polyhedron having twelve faces.

dodecahedron

Between parallel lines: the length of any segment having one endpoint on each line and perpendicular to both.

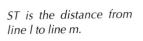

ST is the distance from line l to line m.

Dynamic Parallelogram Any parallelogram whose sides are in the same ratio as one of the Dynamic Rectangles.

divergency In phyllotaxis, the ratio t/n, where *t* is the number of complete turns made around the stem and *n* is the number of leaves passed when travelling from one leaf to another leaf situated directly over it.

For any species of plant, the divergency is a constant.

Dynamic Rectangles A family of rectangles derived from the square through the use of diagonals. If the width is 1, then the lengths of the rectangles are given by square roots of the natural numbers.

The Dynamic Rectangles are sometimes called the Euclidean Series, sometimes root rectangles.

Dynamic Spiral Any logarithmic spiral which uses as its framework one of the Dynamic Rectangles.

Divine Proportion The proportion derived from the division of a line segment into two segments such that the ratio of the whole segment to the longer part is the same as the ratio of the longer part to the shorter part.

A Dynamic Spiral is named by its related Dynamic Rectangle. eg. The spiral created within a √2 Rectangle would be named a √2 Spiral.

E

edge In a polyhedron, the common side between two faces.

In the cube there are twelve edges joining the eight vertices.

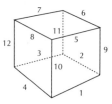

Egyptian Triangle See Triangle of Price.

equiangular Having angles of the same measure.

equiangular triangle

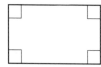

equiangular quadrilateral

equiangular spiral See spiral (logarithmic spiral).

equilateral Having sides of the same measure.

equilateral triangle

equilateral hexagon

escribe To draw around.

Euclidean Series See Dynamic Rectangles.

even number A whole number that is divisible by two.

eg. 2, 4, 6, 8, 10, . . .

If n is any whole number, 2n will always designate an even number.

exponent A number that indicates the number of times a quantity is to be used as a factor.

In the expression a^4, 4 is the exponent.
$a^4 = (a)(a)(a)(a)$.

exterior *Of an angle or a closed curve:* The set of all points in the plane that do not lie on the figure or in the interior of the figure.

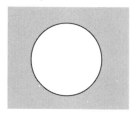

The shaded portion of the plane represents the exterior of the figure.

exterior angle of a polygon An angle formed by a side and the extension of an adjacent side in a polygon.

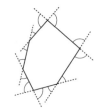

Note: there are two exterior angles at each vertex.

extremes See proportion.

F

face A polygonal shape, together with its interior, that forms part of the surface of a polyhedron.

The shaded portion of the diagram represents one of the six faces of the cube.

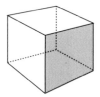

Fibonacci Angle See Ideal Angle.

Fibonacci sequence The summation sequence whose terms are: 1, 1,

2, 3, 5, 8, 13, 21, 34, 55, . . .

G

geometric sequence A sequence in which each term is gotten by multiplying the preceding term by a constant quantity.

If r is the constant, the nth term, s_n, is equal to rs_{n-1}.

eg. 2, 4, 8, 16, 32 . . . is a geometric sequence where each term is multiplied by 2 to produce the following term.

gnomon Any figure that when added to a given figure results in a figure similar to the original.

In each figure, the shaded portion is a gnomon for the unshaded triangle.

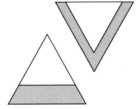

Golden Cut The point on a line segment that divides the segment into two segments whose lengths are in the Golden Ratio.

C is the Golden Cut of $\overline{AB}$.

Golden Mean Same as Golden Cut.

Golden Ratio $\frac{1+\sqrt{5}}{2}$, which is the ratio of approximately 1.61803 to 1, derived from the Divine Proportion.

The Golden Ratio is known by the Greek letter, Ø (phi).

Golden Rectangle Any rectangle whose sides are in the ratio of Ø to 1, where, to five decimal places, Ø=1.61803.

Golden Section Same as Golden Cut.

Golden Sequence The sequence whose terms are the successive powers of Ø: Ø, Ø², Ø³, Ø⁴, Ø⁵, ...

Golden Spiral The logarithmic spiral formed within the pattern of whirling squares in a Golden Rectangle.

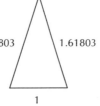

Golden Triangle Any isosceles triangle in which the ratio of a leg to the base is Ø to 1, where, to five decimal places, Ø=1.61803.

Also called the Triangle of the Pentalpha or the Sublime Triangle.

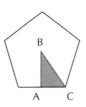

grid A repeating pattern superimposed on the plane consisting of lines and/or circles.

H

Harmonic Triangle The right triangle derived from the radius and apothem of a regular pentagon.

The shaded portion represents an Harmonic Triangle.

helix A spiral-like curve formed by wrapping a cylinder uniformly in a single layer.

The thread of a metal screw is a helix.

heptagon See septagon.

hexagon A polygon having six sides.

hexagons

hue The name of a particular color.

hypotenuse In a right triangle, the side opposite the right angle.

In △ABC, $\overline{AB}$ is the hypotenuse.

hypothesis A conjecture that requires testing for verification.

I

icosagon A polygon having 20 sides.

icosagon

icosahedron A polyhedron having 20 faces.

icosahedron

Ideal Angle The angle measuring about 137.5°, related to plant divergency, which would provide maximum exposure to sunlight as well as minimum overlapping of leaves.

included angle The angle formed by two specified adjacent sides of a polygon.

∠A is included between $\overline{AB}$ and $\overline{AC}$.

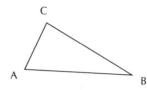

∠B is included between $\overline{AB}$ and $\overline{BC}$.
∠C is included between $\overline{AC}$ and $\overline{BC}$.

included side In a polygon, the common side between two specified angles.

$\overline{AC}$ *is included between ∠A and ∠C.*
$\overline{AB}$ *is included between ∠A and ∠B.*
$\overline{BC}$ *is included between ∠B and ∠C.*

incommensurable Not able to be measured with a common unit.

induction See inductive reasoning.

inductive reasoning A method of reasoning that allows a generalization to be made from observing specific cases.

Inductive reasoning can be likened to making educated guesses, and, unlike deductive reasoning, can lead to a false conclusion. This is the form of reasoning used in the Scientific Method.

inscribed angle An angle whose vertex lies on a circle and each of whose sides intersects the circle.

∠CAB is inscribed in circle P.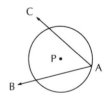

inscribed circle A circle lying in the interior of a polygon such that each side of the polygon is tangent to the circle.

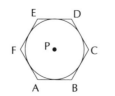

Circle P is inscribed in ABCDEF.

inscribed polygon A polygon whose vertices lie on a circle.

ABCDE is inscribed in circle P.

integer Any member of the following infinite set of numbers:

. . . -3, -2, -1, 0, 1, 2, 3, . . .

interior *Of an angle:* The set of all points not on an angle having the property that if any two are joined by a line segment, the segment will not intersect the angle.

The shaded portions represent the interiors of angles ABC and DEF, respectively.

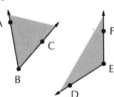

Of closed curve: The set of all points in the plane that are enclosed by the curve.

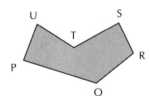

The shaded portion represents the interior of polygon PQRSTU.

interior angle *Of a polygon:* An angle formed by adjacent sides of a polygon.

The number of interior angles is the same as the number of sides or vertices.

intersect To have at least one point in common.

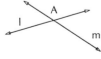

Lines l and m intersect at point A, and A is called the point of intersection.

irrational number Any number that cannot be expressed as the ratio of two integers.

isosceles trapezoid A trapezoid in which the nonparallel sides are congruent.

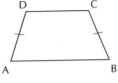

The nonparallel sides are called the legs of the trapezoid.

ABCD is an isosceles trapezoid with legs $\overline{AD}$ and $\overline{BC}$.

isosceles triangle A triangle having a pair of congruent sides.

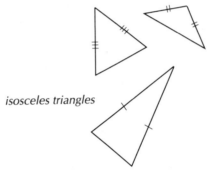

isosceles triangles

L

leg(s) *In an isosceles trapezoid:* the nonparallel sides.

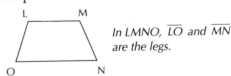

In LMNO, $\overline{LO}$ and $\overline{MN}$ are the legs.

In an isosceles triangle: the congruent sides.

In △PQR, $\overline{PQ}$ and $\overline{PR}$ are the legs.

In a right triangle: the sides that form the right angle.

In △ABC, $\overline{AC}$ and $\overline{CB}$ are the legs.

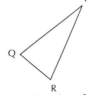

line An undefined term in geometry. It has the properties of infinite length, continuity, and no width. It is a straight curve.

A line is named by any two of its points or by a lower case letter.

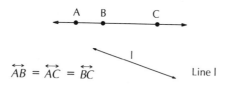

$\overleftrightarrow{AB} = \overleftrightarrow{AC} = \overleftrightarrow{BC}$ Line l

line segment The figure formed by two points on a line and all the points in between those two points.

The two points are called the endpoints of the segment, and a line segment is named by its endpoints.

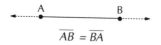

$\overline{AB} = \overline{BA}$

logarithmic spiral See spiral.

M

mean proportion A proportion in which the means are equal.

$\frac{a}{b} = \frac{b}{c}$ *is a mean proportion.*

means See proportion.

median A line segment that joins a vertex of a triangle to the midpoint of the opposite side.

Every triangle has three medians.

medians $\overline{AF}$, $\overline{CE}$, *and* $\overline{BD}$

midpoint The point that divides a line segment into two congruent segments.

R is the midpoint of $\overline{PQ}$.

minute One-sixtieth of a degree in angle measure.

46°17' reads 46 degrees, 17 minutes. 60' = 1°.

monochromatic color scheme A color scheme that uses one hue and the light and dark values thereof.

monohedral tiling A tiling that has a single prototile.

morphology The study of form.

N

natural numbers The counting numbers.

The set of natural numbers is named by the letter N. N = {1,2,3, ...}.

net A bounded portion of a grid or tiling used to define a particular figure.

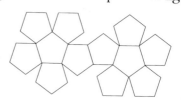

Net for the dodecahedron.

nonagon A polygon having nine sides.

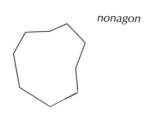

nonagon

noncollinear Not lying on the same line.

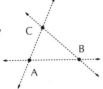

Points A, B, and C are noncollinear.

nonconsecutive Not in order or in sequence.
See consecutive for more information.

numerator In a fraction, the number written above the line.

O

obtuse angle An angle that measures between 90° and 180°.

obtuse angles

obtuse triangle A triangle having one obtuse angle.

obtuse triangles

octagon A polygon having eight sides.

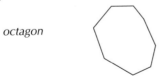

octagon

octahedron A polyhedron having eight faces.

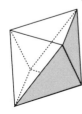

octahedron

odd number A whole number that is not divisible by two.

eg. 1, 3, 5, 7, 9 . . .
If n is any whole number, 2n+1 or
2n-1 will always designate an odd number.

opposite rays The two rays determined by a single point on a line that travel in different directions. Two rays are opposite rays if they are collinear and have only their endpoint in common.

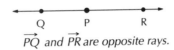

$\overrightarrow{PQ}$ *and* $\overrightarrow{PR}$ *are opposite rays.*

P

palindrome A word or number that reads the same from either end.

12321 and pop are palindromes.

parallel *Lines:* two or more lines that lie in the same plane and have no points in common.

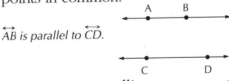

$\overleftrightarrow{AB}$ *is parallel to* $\overleftrightarrow{CD}$.

Segments: noncollinear segments which lie on parallel lines.

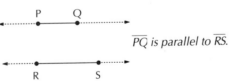

$\overline{PQ}$ *is parallel to* $\overline{RS}$.

parallelogram A quadrilateral in which opposite sides are parallel.

ABCD is a parallelogram.
See Appendix B for properties of a parallelogram.

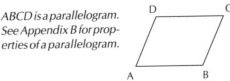

pattern A repetition of units in an ordered sequence.

pentacle See pentagram.

pentagon A polygon having five sides.

pentagon

pentagram The star shaped figure obtained by drawing all the diagonals of a regular pentagon.

Also called a pentalpha, pentacle, or pentangle.

The solid segments represent the pentagram. The dotted segments represent the corresponding regular pentagon.

pentalpha See pentagram.

pentangle See pentagram.

perimeter The distance around a figure.

perpendicular *Lines:* lines that meet to form right angles.

$$\overleftrightarrow{OP} \perp \overleftrightarrow{OQ}.$$

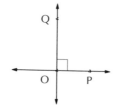

Segments: noncollinear segments that lie on perpendicular lines.

$$\overline{AB} \perp \overline{CD}.$$

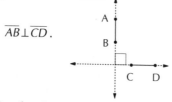

perpendicular bisector See bisector, perpendicular.

phi The number naming the ratio that appears in the Divine Proportion (approximately 1.61803). The Greek letter phi (Ø) is used in this text, but the letter tau (τ) is also used by mathe-

maticians to name this ratio.

Phidian spiral The logarithmic spiral whose successive whorls cut any radius vector into Ø-proportions.

Ø-Family Rectangles The group of rectangles related to the Golden Rectangle or generated by Ø. They are: the square, the $\sqrt{\emptyset}$ Rectangle, the Golden Rectangle, the $\sqrt{4}$ Rectangle, the $\sqrt{5}$ Rectangle and the Ø+1 Rectangle.

Ø-Family Spirals The logarithmic spirals that can be developed within the Ø-Family Rectangles.

Construction methods may be found in Edward B. Edwards' book, Patterns and Design with Dynamic Symmetry.

phyllotaxis The distribution or arrangement of leaves or buds on a stem, or seeds in a flower head.

plane An undefined term in Geometry. A plane is often thought of as "a slice of space" having infinite length and width but no thickness.

Platonic Solids The five regular polyhedra: the tetrahedron, the cube, the octahedron, the dodecahedron, and the icosahedron.

point An undefined term in Geometry. A point is often thought of as a location in space having no dimensions.

pole *Of a polar graph:* the point corresponding to the origin.
Of a spiral: the innermost point of the spiral.

polygon A simple closed curve composed of line segments each intersecting exactly two others, one at each endpoint.

The line segments are called the sides of the polygon and the points of intersection are called the vertices of the polygon. A polygon is named by its vertices read consecutively in either a clockwise or counterclockwise direction.

This polygon could be named ABCDE. It could also be named DCBAE, among others.

polyhedron A three-dimensional figure formed by polygons which intersect in such a way that vertices and sides are common.

The polygonal figures are called the faces of the polyhedron, the common sides are called the edges of the polyhedron, and the common vertices are called the vertices of the polyhedron.

primary colors In pigments, yellow, red and blue.

These are the essential hues from which all others are derived.

prime number A natural number, different from one, that is divisible only by one and itself.

There are an infinite number of prime numbers. The sequence of primes is as follows: 2, 3, 5, 7, 11, 13, 17, 19, 23, 29, . . .

proportion A pair of ratios set equal to each other.

$$\frac{a}{b} = \frac{c}{d}$$

*In this proportion, **a** and **d** are called the extremes, and **b** and **c** are called the means of the proportion. These words refer only to the position of the terms.*

proportional spiral See spiral (logarithmic spiral).

protractor A tool used for measuring the number of degrees in an angle.

protractor

pyramid A three-dimensional figure set on a polygonal base whose lateral faces are triangles formed by segments joining the vertices of the polygon to a single point (called the vertex) not in the same plane.

If the base is regular and the segment joining its center to the vertex is perpendicular to the plane of the base, the pyramid is a right pyramid.

Pythagorean Theorem $a^2 + b^2 = c^2$, where a and b name the lengths of the legs of a right triangle and c names the length of the hypotenuse.

Pythagorean triple or triad Three integers that have the property that the sum of the squares of two of them equals the square of the third.

eg. 3, 4, 5 is a Pythagorean triple since $3^2 + 4^2 = 5^2$ or 9 + 16 = 25.

Q

quadratic equation An equation of the form $ax^2 + bx + c = 0$, where a, b and c are real numbers and $a \neq 0$.

quadratic formula

$$x = \frac{-b \pm \sqrt{b^2 - 4ac}}{2a}$$

where a, b and c are the coefficients of a quadratic equation $ax^2 + bx + c = 0$.

quadrilateral A four sided polygon.

quadrilaterals

R

radicand The quantity under the radical sign.

eg. in √15, 15 is the radicand.
In √3xy, 3xy is the radicand.

radius *Of a circle:* a line segment that joins the center of a circle to any point on the circle is *a* radius of the circle. The measure of any of these segments (all are congruent) is *the* radius of the circle.

$\overline{OP}$ *is a radius of circle O.*

Of a regular polygon: if a regular polygon is inscribed in a circle, *the* radius of the polygon is the radius of the circumscribed circle. A radius is a line segment from the "center" of the polygon to a vertex.

$\overline{AB}$ *is a radius of BCDEFG.*

radius vector Any ray emanating from the pole of a spiral and intersecting the spiral curve.

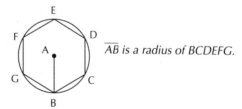

$\overrightarrow{PQ}$ *is a radius vector.*

ratio A comparison of numbers to each other.

The ratio of 2 to 3 may be expressed as 2:3 or simply 2 to 3 or as a fraction 2/3.

rational number Any number that can be put into the form p/q where p and q are integers and q≠0.

ray The figure formed by a point on a line and all the points on the same side of that point.

The point is called the endpoint of the ray, and the ray is named by its endpoint and any other point on the ray.

$\overrightarrow{AB} = \overrightarrow{AC}$.

reciprocal *Of a number:* if *n* is any number then 1/n is its reciprocal.

The product of any number and its reciprocal is always one. (n)(1/n)=1.

Of a rectangle: a similar rectangle cut from the parent rectangle such that its length is the width of the parent rectangle.

ABCD ~ BCEF and BCEF is the reciprocal of ABCD.

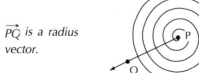

rectangle A parallelogram whose angles are right angles.

ABCD is a rectangle.

Rectangle of Price See √Ø Rectangle.

regular *Polygon:* a polygon having all sides and angles congruent.

Regular polyhedron: A polyhedron whose faces are all congruent regular polygons.

rhombus A parallelogram in which all sides are congruent.

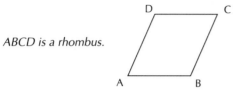

ABCD is a rhombus.

right angle An angle that measures 90°.

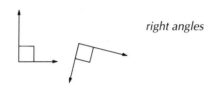

right angles

right triangle A triangle having one right angle.

right triangles

root Same as solution.

root rectangles The rectangles derived from the square such that the width measures one, and the length measures the square root of a natural number. These rectangles are the members of the family of Dynamic Rectangles.

Rope Knotter's Triangle The right triangle whose legs measure 3 and 4 respectively, and whose hypotenuse measures 5 (or 3-4-5 Right Triangle).

It was so named because it was discovered by the ancient Egyptian rope knotters who prepared ropes for purposes of measuring.

S

scalene triangle A triangle having no congruent sides.

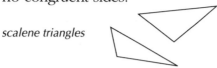

scalene triangles

secant A line that intersects a circle in exactly two points.

Line l is a secant with respect to circle P.

second One-sixtieth of a minute in angle measure.

60" = 1' or 3600" = 1°.

18°32'21" reads 18 degrees, 32 minutes, 21 seconds.

septagon A polygon having seven sides, sometimes called a heptagon.

septagon

sequence An ordered progression of terms.

A sequence may have a finite number of terms, eg. 1, 2, 3, 4, 5, 6; or an infinite number of terms, eg. 1, 1, 2, 3, 5, 8, 13, . . .

If the notation S_1, S_2, S_3, ..., S_n ... is used to denote a sequence, the subscript names the position of the term. eg. S_{10} would be the tenth term.

side *Of an angle:* see angle.

Of a line: either of the half planes formed by a line in the plane.

The shaded portions represent the two half planes. A and B lie on the same side of line l. A and C lie on opposite sides of line l.

Of a polygon: see polygon.

similar Having the same shape.
(Continued next page.)

Two polygons are similar if corresponding angles are congruent and corresponding sides are in proportion.

$\triangle ABC \sim \triangle DEF$

$$\frac{AB}{DE} = \frac{BC}{EF} = \frac{AC}{DF}$$

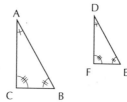

simple closed curve A plane curve that can be traced without lifting pen from paper in such a way that the starting point and ending point are the same and no point is traced more than once.

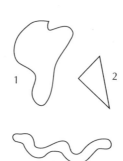

Figs. 1, 2 and 3 are simple closed curves. Figs. 4 and 5 are not.

skew lines Lines that lie in different planes.

solution Any quantity which may be substituted for a variable in an equation to produce a true statement.

eg. 3 is the solution for the equation x + 1 = 4.

spiral A curve which moves around a central point, called the pole, while continually increasing its distance from the pole.

Archimedean spiral: A spiral whose successive whorls are spaced at equal intervals along any radius vector.

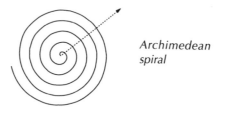

Archimedean spiral

logarithmic spiral: A spiral whose successive whorls form congruent angles with any radius vector and are spaced in geometric progression along the radius vector.

Also known as equiangular spiral, proportional spiral, Spira Mirabilis.

logarithmic spiral

Spira Mirabilis See spiral (logarithmic spiral).

square *Noun:* A rectangle in which all of the sides are congruent.

ABCD is a square.

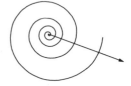

Verb: To multiply a quantity by itself.

The number 2 used as an exponent indicates the operation of squaring. eg. $a^2 = (a)(a)$.

square root A number that when multiplied by itself produces a given number.

The operation of taking the positive square root of a number is indicated by a radical sign and the given number, under the sign, is called the radicand.
eg. $\sqrt{16}$ = 4 since 4 x 4 = 16.

√Ø Rectangle (Square Root of Phi Rectangle): Any rectangle in which the ratio of the length to the width is √Ø to 1, or approximately 1.271.

straight angle An angle measuring 180°.

∠A is a straight angle, so called because its sides form a line.

Sublime Triangle See Golden Triangle.

summation sequence A sequence whose terms are arranged such that any term is the sum of the two preceding terms.

$u_1, u_2, u_3, ..., u_n, ...$ is a summation sequence if $u_i + u_{i+1} = u_{i+2}$ for all integers i.

supplementary angles Two angles the sum of whose measures is 180°.

The angles are said to be supplements of each other.

∠*ABC and* ∠*CBD are supplementary.*

T

tangent A line lying in the same plane as a circle that intersects the circle in exactly one point.

Line m is tangent to circle P at A. A is called the point of tangency.

tetrahedron A polyhedron having four triangular faces.

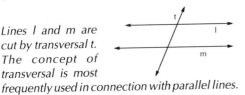 *tetrahedron*

transversal A line that intersects two or more other lines.

Lines l and m are cut by transversal t. The concept of transversal is most frequently used in connection with parallel lines.

trapezoid A quadrilateral with exactly one pair of parallel sides.

The parallel sides are called the bases of the trapezoid and the angles including either base are called the base angles. There are two pairs of base angles in a trapezoid.

trapezoids

triangle A polygon having three sides.

Individual types of triangles are listed under their respective names.

Triangle of Price Any right triangle whose sides are in the ratio of 1; $\sqrt{\varnothing}$: $\varnothing$.

Also called the Egyptian Triangle.

Triangle of the Pentalpha See Golden Triangle.

triangulation The process of dividing an irregular shape into triangular shapes for purposes of finding area.

U

unit measure The length taken to represent the number one in a given set of circumstances.

V

vertex *Of an angle:* see angle.

Of a polygon: see polygon.

volume The measure in cubic units of any three-dimensional object.

W

Whirling Square Rectangle Another name for the Golden Rectangle.

whorl A circular or spiral arrangement of like things such as leaves, petals, hair, etc.

Index

This page:
Chambered Nautilus. Graham Boles.
Top:
Age 6. Ball point pen on paper.
Bottom:
Age 16. Technical pen and ink on paper.